A Sampler of Useful Computational Tools for Applied Geometry, Computer Graphics, and Image Processing

A Sampler of Useful Computational Tools for Applied Geometry, Computer Graphics, and Image Processing

Daniel Cohen-Or (Editor)
Chen Greif
Tao Ju
Niloy J. Mitra
Ariel Shamir
Olga Sorkine-Hornung
Hao (Richard) Zhang

CRC Press
Taylor & Francis Group
Boca Raton London New York

CRC Press is an imprint of the
Taylor & Francis Group, an **informa** business

A CHAPMAN & HALL BOOK

CRC Press
Taylor & Francis Group
6000 Broken Sound Parkway NW, Suite 300
Boca Raton, FL 33487-2742

First issued in paperback 2020

ISBN-13: 978-1-4987-0628-5 (hbk)
ISBN-13: 978-0-367-65878-6 (pbk)

Library of Congress Cataloging-in-Publication Data

A sampler of useful computational tools for applied geometry, computer graphics, and image
    processing / Daniel Cohen-Or, editor ; Chen Greif, Tao Ju, Niloy J. Mitra, Ariel Shamir, Olga
    Sorkine-Hornung, Hao (Richard) Zhang.
        pages cm
    Includes bibliographical references and index.
    ISBN 978-1-4987-0628-5 (alk. paper)
    1. Engineering mathematics. 2. Geometry. 3. Image processing--Mathematics. I. Cohen-Or,
Daniel, editor. II. Greif, Chen, 1965-

TA330.S26 2015
510--dc23   2014044036

Visit the Taylor & Francis Web site at
http://www.taylorandfrancis.com

and the CRC Press Web site at
http://www.crcpress.com

About the Book

Many important topics in applied geometry cannot be solved efficiently without using mathematical tools. This book presents, in a rather light way, some mathematical tools that are useful in many domains, such as computer graphics, image processing, computer vision, digital geometry processing and geometry in general.

What the book includes

This book presents a collection of mathematical tools that correspond to a broad spectrum of applied mathematics and computer science. The book consists of thirteen chapters, each of which covers one topic. The topics are deliberately meant to be uncoupled and can be read or taught in any order. However, the first two chapters discuss more fundamental tools, and serve either to recall or to introduce essential elementary notions from analytical geometry and linear algebra. The remaining chapters cover a wide range of topics, from matrix decomposition to curvature analysis, principle component analysis to dimensionality reduction and more. The full list is presented below.

The presentation style

This book has a special presentation style that gives up excessive rigor and completeness, but strives to be useable, effective and well motivated. The mathematical tools are described and presented as solutions to specific applied problems such as image alignment, surface approximation, compression or image manipulation. The book is meant to be practical and handy, and we hope that reading it will be an enjoyable experience.

To whom the book will appeal

By avoiding detailed proofs and analysis, the book will appeal to those who wish to enrich their problem-solving arsenal. The book is ideal for people who do not have a very deep academic background in mathematics, and yet wish to use mathematics for work or research.

The book is structured for use as a one-semester, intermediate-level course in computer science, both in terms of its length and the required level of knowledge. In fact, the idea for this book was born out of a course that we have been teaching and improving over several years. Each chapter represents the material taught in a weekly class. The course's main objective was to quickly introduce the students to knowledge and tools useful in their advanced studies in a visual computing or applied geometry field. As such, the course primarily served faculty and senior researchers who wanted their students to acquire this arsenal. However, the book is also perfectly suited for individuals who wish to learn how to solve non-trivial geometric problems.

Is it a text book?

Not by standard definitions, but it can certainly accompany a course, as explained above. Each chapter can be taught within one session, and further reading material is indicated within the chapters, when appropriate.

Book Contents

Olga Sorkine-Hornung and Daniel Cohen-Or

In the first chapter, we will familiarize ourselves with some basic geometric tools and see how we can put them to practical use to solve several geometric problems. Instead of describing the tools directly, we do it through an inter- esting discussion of two possible ways to approach the geometric problem at hand: we can employ our geometric intuition and use geometric reasoning, or we can directly formalize everything and employ our algebraic skills to write 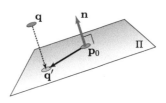 down and solve some equations. The discussion leads to a pre- sentation of linear geometric elements (points, lines, planes), and the means to manipulate them in common geometric applications that we encounter, such as distances, transformations, projections and more.

Daniel Cohen-Or, Olga Sorkine-Hornung and Chen Greif

In this chapter, we will review basic linear algebra notions that we learned in a basic linear algebra course, including vector spaces, or- thogonal bases, subspaces, eigenvalues and eigenvectors. However,  our main goal here is to convince the readers that these notions are really useful. Furthermore, we will see the close relation between linear algebra and geometry. The chapter will be driven by an important tool called singular value decomposition (SVD), to which we will devote a separate full chapter. To understand what an SVD is, we first need to understand the notions of bases, eigenvectors, and eigenvalues and to refresh some fundamentals of linear algebra with examples in geometric context.

When dealing with real-world data, simple patterns can often be submerged in noise and outliers. In this chapter, we will learn about basic data fitting using the least-squares method, first starting with simple line fitting before moving on to fitting low-order polynomials. Beyond robustness to noise, we will also learn how to handle outliers and look at basic robust statics.

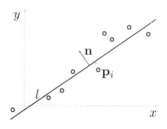

In this chapter, we introduce two related tools from linear algebra that have become true workhorses in countless areas of science: principal component analysis (PCA) and singular value decomposition (SVD). These tools are extremely useful in geometric

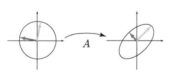

modeling, computer vision, image processing, computer graphics, machine learning and many other applications. We will see how to decompose a matrix into several factors that are easy to analyze and reveal important properties of the matrix and hence the data, or the problem in which the matrix arises. As in the whole book, the presentation is rather light, emphasizing the main principles without excessive rigor.

Hao (Richard) Zhang

The use of signal transforms, such as the discrete Fourier or cosine transforms, is a classic topic in image and signal processing. In this chapter, we will learn how such transforms can be formulated and applied to the  processing of 2D and 3D geometric shapes. The key concept to take away is the use of eigenvectors of discrete Laplacian operators as basis vectors to define spectral transforms for geometry. We will show how the Laplacian operators can be defined for 2D and 3D shapes, as well as a few applications of spectral transforms including geometry smoothing, enhancement and compression.

Chen Greif

In the solution of problems discussed in this book, a frequent task that arises is the need to solve a linear system. Understanding the properties of the matrix associated with the linear 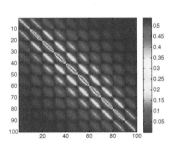 system is critical for guaranteeing speed and accuracy of the solution procedure. In this chapter, we provide an overview of linear system solvers. We describe direct methods and iterative methods, and discuss important criteria for the selection of a solution method, such as sparsity and positive definiteness. Important notions such as pivoting and preconditioning are explained, and a recipe is provided that helps in determining which solver should be used.

Daniel Cohen-Or and Gil Hoffer

In this chapter, we make use of the well-known equations of Laplace and Poisson. The two equations have an extremely simple form, and they are very useful in many diverse branches of mathematical physics. However, in this chapter, we will interpret them in the context of image processing. We will show some interesting image editing and geometric problems and how they can be solved by simple means using these equations.

Niloy J. Mitra and Daniel Cohen-Or

Local surface details, e.g., how "flat" a surface is locally, carry important information about the underlying object. Such information is critical for many applications in geometry processing, ranging from surface meshing, shape matching, surface reconstruction, scan alignment and detail-preserving deformation, to name only a few. In this chapter, we will cover the basics of differential geometry, particularly focusing on curvature estimates with some illustrative examples as an aid to geometry processing tasks.

  Hao (Richard) Zhang and Daniel Cohen-Or

In this chapter, we will learn the concept, usefulness, and execution of dimensionality reduction. Generally speaking, we will
seek to reduce the dimensionality of a given data set, mapping high-dimensional data into a lower-dimensional space to facilitate visualization, processing, or inference. We will present and discuss only a sample of dimensionality reduction techniques and illustrate them using visually intuitive examples, including face recognition, surface flattening and pose normalization of 3D shapes.

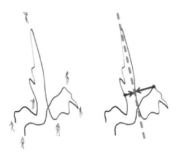

  Tao Ju

In this chapter, we visit the classical mathematical problem of obtaining a continuous function over a spatial domain from data

at a few sample locations. The problem comes up in various geometric modeling scenarios, a good example of which is surface reconstruction. The chapter will eventually introduce the very useful radial basis functions (RBFs) as a smooth and efficient solution to the interpolation problem. However, to understand their usefulness, the chapter will go
through a succession of methods with increasing sophistication, including piecewise linear interpolation and Shepherd's method.

Chapter 11: Topology: How Are Objects Connected? 163
Niloy J. Mitra

In Chapter 8, we learned about local differential analysis of surfaces. In this chapter, we focus on global aspects. We will learn about what is meant by orientable surfaces or manifold surfaces. Most importantly, we will learn about the Euler characteristic, which links local curvature properties to global connectivity constraints, and comes up in a surprising range of applications.

Chapter 12: Graphs and Images.........................177
Ariel Shamir

Graphs play an important role in many computer science fields and are also extensively used in imaging and graphics. This chapter concentrates on image processing and demonstrates how images can be represented by a graph. This allows translating problems of analysis and manipulation of images to well-known graph algorithms. Specifically, we will show how segmentation of images can be solved using region-growing algorithms such as watershed or partitioning algorithms using graph cuts. We will also

show how intelligently changing the size and aspect ratio of images and video can be solved using dynamic programming or graph cuts.

In this chapter, we will show an example of the usefulness of number theory, or at least one of its known theorems. We will discuss mappings of numbers to a lattice, a problem that has practical applications in systems that require simultaneous, conflict-free access to elements distributed in different memory modules. Such mappings are also called *skewing schemes* since they skew the trivial mapping from element to memory. To understand these mappings, we will visit the notions of relatively prime numbers, and the greatest common divisor (gcd).

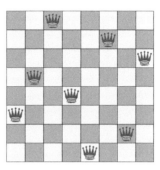

Chapter 1

Analytical Geometry

Olga Sorkine-Hornung and Daniel Cohen-Or

When dealing with geometric problems, we typically deal with 2D and 3D objects. These objects are represented mathematically in the computer. Therefore, the basic requirement is to have a good grasp of the mathematical entities that represent the objects, and be familiar with the tools that enable their manipulation. The basic geometric elements are linear (points, lines, planes), and the typical queries and manipulations we encounter in common geometric applications include distances, transformations, projections and so on.

All these manipulations require performing computations based on the mathematical representation of digital objects, by using recipes from analytical geometry and linear algebra. In this chapter, we will familiarize ourselves with some basic geometric tools and see how we can put them to practical use to solve several geometric problems. However, before we dive into the details, let us discuss how can we approach geometric problems in general.

Solving geometric problems

When we have a geometric problem at hand, there are two possible ways to look at it: we can employ our geometric intuition and use geometric reasoning first, or we can directly formalize everything and employ our algebraic skills to write down and solve some equations. We will often try to use geometric understanding

1

first, which will possibly simplify the problem, and then use the algebra. Sometimes, when the problem is complex, it is hard to find intuitive geometric insights, and then we have to rely on algebraic tools to help us.

Let us look at our first example of a geometric problem and discuss the two possible ways to attack it.

Problem: *Given two straight lines in the 3D space, compute the distance between the two lines.*

Let us first try and think of the problem in pure geometric terms, without equations. The first observation is that the distance between two lines in 3D is the length of the shortest segment connecting them. This segment is perpendicular to both lines. We can

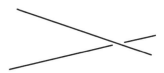

Figure 1.1: Two lines in 3D space, in general configuration.

look at the problem of finding that segment to try and simplify the distance problem. Indeed, imagine that we are looking at the installation of our two lines, so that the direction of our sight coincides with one of the lines, say, the first line. Then we will not see the first line at all, instead we will see just a single point! So the (hypothetical) image of the scene in our eye will be a line and a point, where the line is actually the projection image of the second line onto the plane that is perpendicular to the first line (see Figure 1.2). Thus, our problem becomes simpler: now we need to compute the distance between a point and a line.

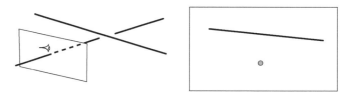

Figure 1.2: If we look at the scene in the direction of one line, that line becomes a single point.

We can simplify even further by using the same geometric argument: let us now imagine that we are looking at the result of the previous mental exercise (a point and a projected line), and our line of sight coincides with the line again (see Figure 1.3). As before, in our mental image, the line turns into a point, and we are left with the task of computing the distance between two points!

Figure 1.3: If we look at the scene in the direction of the line, that line becomes a single point.

How do we get the actual distance? After having envisioned the required manipulations of the scene, we need to formalize all the objects and transformations. Namely, we need to take the mathematical representations of the two lines, perform the two orthogonal projections, and finally, compute the distance between the two resulting points. As you can see, there is no way to completely escape the math; however, the geometric approach provides us with a very convenient recipe that is easy to program, because it consists of simple building blocks. If we have the basic geometric routines implemented (like projections of lines and points onto a plane), all we need to do is call these routines with the appropriate input. Another advantage of having a geometric intuition backing us up is the ability to quickly find mistakes and debug the program: since we have a good idea of what the result should look like, we can render the results of the algorithm step by step and quickly identify what went wrong.

An alternative way to approach the same problem is directly through the mathematical formulations. In this case, we almost skip the geometric observations, apart from the basic fact that the shortest distance between two lines should be achieved between two points on those lines, such that the segment connecting these two points is perpendicular to both lines. We take the mathematical representation of the two lines, and formulate and solve a system of equations that will give us the two points that achieve the shortest distance:

$$\langle \mathbf{p}_1 - \mathbf{p}_2, \mathbf{v}_1 \rangle + s\|\mathbf{v}_1\|^2 - t\langle \mathbf{v}_1, \mathbf{v}_2 \rangle = 0 \,,$$
$$\langle \mathbf{p}_1 - \mathbf{p}_2, \mathbf{v}_2 \rangle + s\langle \mathbf{v}_1, \mathbf{v}_2 \rangle - t\|\mathbf{v}_1\|^2 = 0 \,.$$

The meaning of these equations, in particular, the scalar product operator $\langle \cdot, \cdot \rangle$, will be explained later in this chapter; in the meantime we will just mention that these equations are a direct result of writing the perpendicularity requirement mathematically. It is quite straightforward to write such an encoding of the problem once we have good command of the analytic tools; however, such an approach might be harder to program and debug because it is less intuitive.

Let us now look at a second example of a geometric problem and discuss the two possible ways to approach it.

Problem: *Compute the intersection between a ray and sphere. The sphere is of radius r and centered at $\mathbf{c}$, and the ray emanates from point $\mathbf{p}_0$ in direction $\mathbf{v}$.*

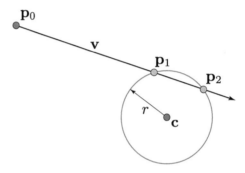

Figure 1.4: A ray intersecting a sphere at two points.

Here as well, two approaches can be considered, namely, one purely analytical and one geometric. It should be noted that in many applications, this problem is often more specific, for example in ray tracing: one is interested only in the *first* intersection of the ray with the sphere, or only in a binary answer, that is, whether or not the ray intersects the sphere.

The pure analytical solution is straightforward (see Figure 1.4): we take the parametric representation of the ray, $\mathbf{p}(t) = \mathbf{p}_0 + t\mathbf{v}$, and the algebraic representation of the sphere, $\|\mathbf{p} - \mathbf{c}\|^2 - r^2 = 0$, and simply substitute for $\mathbf{p}(t)$ to get $\|\mathbf{p}_0 + t\mathbf{v} - \mathbf{c}\|^2 - r^2 = 0$. This is a quadratic equation in t. Solving for t yields two, one, or no solutions. If the ray intersects the sphere, typically there are two solutions, and only one if the ray is tangent to the sphere. However, with this approach we simply plug the scene parameters into the quadratic equation and solve.

A geometric approach requires us to build the setting and observe the relations among the geometric entities involved. This opens opportunities to reach the solution by simple incremental steps, allowing early termination of the computation in case the ray does not meet the sphere at all. Observe Figure 1.5. Let $\mathbf{b}$ denote the line segment connecting the ray source $\mathbf{p}_0$ and the sphere center $\mathbf{c}$: $\mathbf{b} = \mathbf{c} - \mathbf{p}_0$. Let $\mathbf{m}$ denote the point along the ray that is closest to $\mathbf{c}$, and d the distance between $\mathbf{c}$ and the ray, which is given by $\|\mathbf{c} - \mathbf{m}\|$. Note that the minimal distance between the ray and $\mathbf{c}$ is when the vectors $\mathbf{m} - \mathbf{p}_0$ and $\mathbf{m} - \mathbf{c}$ are perpendicular.

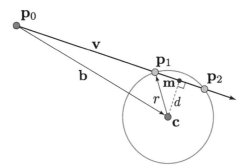

Figure 1.5: A ray intersecting a sphere at two points. The point $\mathbf{m}$ is the closest point along the ray to the center $\mathbf{c}$ of the sphere.

The distance along the ray from $\mathbf{p}_0$ to $\mathbf{m}$ is given by the dot product (if $\mathbf{v}$ is a unit vector):

$$t_b = \langle \mathbf{b}, \mathbf{v} \rangle. \tag{1.1}$$

Later on in this chapter we will refresh the memory of those who have forgotten the meaning of a dot product. Having t_b in hand allows us to terminate the computation if $t_b < 0$, which means that the ray travels in the opposite direction and does not meet the sphere. This is called a *trivial reject*. Moreover, knowing t_b also allows us to easily test whether $d = \|\mathbf{m} - \mathbf{c}\|$ is larger than the sphere radius, which again means that no intersection occurs. Applying the Pythagorean formula, we can calculate $d^2 = \|\mathbf{b}\|^2 - t_b{}^2$ and compare to r^2; since we are interested in a binary test, we do not need to explicitly compute d, avoiding the expensive calculation of a square root.

Only if the above two tests are positive do we compute the actual intersection point using a square root: $t_* = \sqrt{r^2 - d^2}$. The two intersection points $\mathbf{p}_1$ and $\mathbf{p}_2$ are given (in terms of the ray parameter t) by $t_b \pm t_*$, and the closest intersection is the smaller value of $t_b \pm t_*$.

The early termination is very effective, since in most applications the ray does not meet the given sphere in the majority of cases. In other cases, only a binary test is required, for example, if the sphere is used as a bounding volume, where the ray–sphere intersection is just meant to test whether the ray should go farther and intersect the object bounded by the sphere. Clearly, in such situations, applying the analytical formula for ray–sphere intersection is overly expensive.

Analytical approach

As mentioned, not every problem can be approached from a purely geometric point of view; in complex situations we have to rely on algebraic tools to help us. And in any case, even if we first formulate the problem in geometric terms, eventually we would have to deal with algebraic formalisms. Therefore, in the following we review the basic notions from analytical geometry (and linear algebra in the next chapter), and see how to put them to use in solving various geometric problems, in particular one of those mentioned above.

We start with a recap of the basic geometric entities: points and vectors. We assume the reader has heard of these notions somewhere, along with some basic algebra, and therefore we do not attempt to teach this from scratch, but rather recall the basic definitions, facts and notation.

Points and vectors. Points specify locations in space. We denote points by bold letters: $\mathbf{p}, \mathbf{q}$, etc. Vectors specify a direction and magnitude, like velocity. They do not have a particular location in space. We also denote vectors by bold letters: $\mathbf{v}, \mathbf{w}$, etc. It should always be clear from the context whether we are talking about points or vectors; in the following we describe some rules for combining them. It is important to realize that points and vectors are not the same, even though they have the same mathematical representation using coordinates.

We can combine points and vectors using arithmetical operators. Adding a vector $\mathbf{v}$ to a point $\mathbf{p}$ results in another point $\mathbf{q}$, which is the translation of $\mathbf{p}$ in the direction of vector $\mathbf{v}$, going a distance that amounts to the magnitude of $\mathbf{v}$ (see Figure 1.6 (left)).

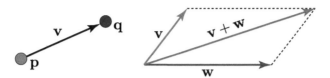

Figure 1.6: Adding a vector to a point (left). Adding two vectors (right).

We can also add vectors, the result being another vector. The vector addition is defined via the famous parallelogram rule, best described visually (see Figure 1.6 (right)). We can also subtract vectors, using the vector negation operation:

$$\mathbf{v} - \mathbf{w} = \mathbf{v} + (-\mathbf{w}).$$

We know how to add two vectors using the parallelogram rule; the negation of a vector, $-\mathbf{w}$, is simply reversing the direction of $\mathbf{w}$ (the magnitude remains the same).

We can subtract points from one another; the result of $\mathbf{q} - \mathbf{p}$ is a vector! It is the vector whose magnitude is the distance between the two points and whose direction is the direction of going from $\mathbf{p}$ to $\mathbf{q}$ (see Figure 1.7).

Figure 1.7: Subtracting points.

The addition of points is not defined! Since points are locations, adding them makes no sense. However, we can establish a one-to-one mapping between points and vectors, and we can add a pair of vectors that corresponds to a pair of points. To define the mapping between vectors and points, we need one reference location in space, which we will denote by $\mathbf{o}$ (it is also called *origin*). Given an origin point $\mathbf{o}$, we define the correspondence by

$$
\begin{aligned}
\text{Point to vector:} \quad & \mathbf{p} \to \mathbf{p} - \mathbf{o}, \\
\text{Vector to point:} \quad & \mathbf{v} \to \mathbf{o} + \mathbf{v}.
\end{aligned}
$$

Scalar product. The scalar product is a fundamental operator on two vectors, providing us with a scalar quantity. It allows us to conveniently compute projections and is widely used in all our computations. The scalar product operator is also called the "dot product" or "inner product."

The definition of the scalar product is as follows. Given two vectors $\mathbf{v}$ and $\mathbf{w}$ and the angle θ between them,

$$\langle \mathbf{v}, \mathbf{w} \rangle = \|\mathbf{v}\| \cdot \|\mathbf{w}\| \cdot \cos\theta .$$

Observe Figure 1.8, where ℓ denotes the length of the projection of $\mathbf{w}$ onto $\mathbf{v}$. From basic trigonometry, we have $\cos\theta = \frac{\ell}{\|\mathbf{w}\|}$, which leads to

$$\ell = \frac{\langle \mathbf{v}, \mathbf{w} \rangle}{\|\mathbf{w}\|},$$

a very useful application of the dot product operator, as we demonstrated in Equation (1.1).

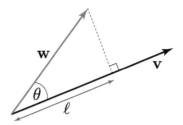

Figure 1.8: Computing $\cos\theta = \frac{\ell}{\|\mathbf{w}\|}$.

The dot product of two vectors $\langle \mathbf{v}, \mathbf{w} \rangle$ can be computed explicitly given a coordinate representation of the vectors (more on this later in Chapter 2): given $\mathbf{v} = (x_\mathbf{v}, y_\mathbf{v})$ and $\mathbf{w} = (x_\mathbf{w}, y_\mathbf{w})$, we have $\langle \mathbf{v}, \mathbf{w} \rangle = x_\mathbf{v} x_\mathbf{w} + y_\mathbf{v} y_\mathbf{w}$. In 2D we have a useful property that if $\langle \mathbf{v}, \mathbf{w} \rangle = 0$, then the vectors are perpendicular, and we can compute the perpendicular vector of $\mathbf{v} = (x_\mathbf{v}, y_\mathbf{v})$ as $\mathbf{v}^\perp = (-y_\mathbf{v}, x_\mathbf{v})$.

Parametric representation of a line. A parametric representation of a line is given by $l(t) = \mathbf{p} = \mathbf{p}_0 + t\mathbf{v}$, where the line origin is at $\mathbf{p}_0$, where $t = 0$ (see Figure 1.9); the parameter value is any real number $t \in (-\infty, \infty)$. When we refer to a ray, rather than a line, the forward direction of the ray is where $t > 0$.

With a parametric representation of a line, we can derive the distance between a point $\mathbf{q}$ and a line $l(t) = \mathbf{p}_0 + t\mathbf{v}$. As shown in

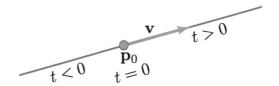

Figure 1.9: Parametric representation of a line.

Figure 1.10, we look for a point $\mathbf{q}'$ such that $\mathbf{q} - \mathbf{q}'$ is perpendicular to $\mathbf{v}$, and then the distance between $\mathbf{q}$ and l is $\mathrm{dist}(\mathbf{q}, l) = \|\mathbf{q} - \mathbf{q}'\|$.

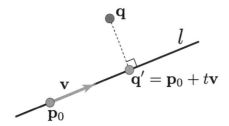

Figure 1.10: The distance between a point and a line.

Let us denote it in a dot product form:

$$\langle \mathbf{q} - \mathbf{q}', \mathbf{v} \rangle = 0 \,,$$
$$\langle \mathbf{q} - (\mathbf{p}_0 + t\mathbf{v}), \mathbf{v} \rangle = 0 \,,$$
$$\langle \mathbf{q} - \mathbf{p}_0, \mathbf{v} \rangle - t\langle \mathbf{v}, \mathbf{v} \rangle = 0 \,.$$

Solving for t:

$$t = \frac{\langle \mathbf{q} - \mathbf{p}_0, \mathbf{v} \rangle}{\|\mathbf{v}\|} \,.$$

Now we can apply the Pythagorean formula to get

$$\mathrm{dist}^2(\mathbf{q}, l) = \|\mathbf{q} - \mathbf{p}_0\|^2 - t^2 \,.$$

Note that the parametric representation of the line is coordinates-independent. The vectors and points $\mathbf{v}$, $\mathbf{p}_0$ and $\mathbf{q}$ can be in 2D or in 3D or in any dimension.

In a similar fashion, we can derive the distance between a point and plane. First, a given plane Π is characterized by its normal $\mathbf{n}$, see Figure 1.11. Given a point $\mathbf{q}$, its distance to the plane is $\|\mathbf{q}' - \mathbf{q}\|$, where $\mathbf{q}'$ is the projection of $\mathbf{q}$ in the direction of the normal vector $\mathbf{n}$ onto the plane. Since $\mathbf{q}' - \mathbf{q}$ is parallel to $\mathbf{n}$,

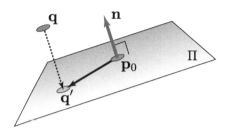

Figure 1.11: The distance between a point and a plane.

$\mathbf{q}' = \mathbf{q} + \alpha\mathbf{n}$, where $\alpha \in \mathbb{R}$. Let $\mathbf{p}_0$ be an arbitrary point on the plane Π, then

$$\langle \mathbf{q}' - \mathbf{p}_0, \mathbf{n} \rangle = 0\,.$$

Substituting $\mathbf{q}'$ we get

$$\langle \mathbf{q} + \alpha\mathbf{n} - \mathbf{p}_0, \mathbf{n} \rangle = 0\,,$$
$$\langle \mathbf{q} - \mathbf{p}_0, \mathbf{n} \rangle + \alpha\langle \mathbf{n}, \mathbf{n} \rangle = 0\,,$$
$$\alpha = \frac{\langle \mathbf{p}_0 - \mathbf{q}, \mathbf{n} \rangle}{\|\mathbf{n}\|^2}\,,$$

and therefore

$$\mathrm{dist}^2(\mathbf{q}, \Pi) = \|\mathbf{q}' - \mathbf{q}\|^2 = \alpha^2\|\mathbf{n}\|^2 = \frac{\langle \mathbf{q} - \mathbf{p}_0, \mathbf{n} \rangle^2}{\|\mathbf{n}\|^2}\,.$$

Note the relation between the distance between a point and a plane to the distance between a point and line. The distance of a point to a plane is no more than a dot product of $\mathbf{q} - \mathbf{p}_0$ and the normal vector. Likewise in 2D, the distance of a point $\mathbf{q}$ to a line in direction $\mathbf{v}$ can be similarly computed by taking $\mathbf{n} = \mathbf{v}^\perp$.

The distance between two lines in 3D.   We are now ready to show the details of the solution to the problem we started with in this chapter (see Figure 1.12). Let l_1 and l_2 be two given arbitrary lines in 3D:

$$l_1(s) = \mathbf{p}_1 + s\mathbf{u}\,,$$
$$l_2(t) = \mathbf{p}_2 + t\mathbf{v}\,.$$

In Figure 1.12, the distance is attained between two points $\mathbf{q}_1$ and $\mathbf{q}_2$ so that $(\mathbf{q}_1 - \mathbf{q}_2)\perp\mathbf{u}$ and $(\mathbf{q}_1 - \mathbf{q}_2)\perp\mathbf{v}$. Thus, we are after the

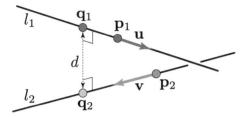

Figure 1.12: The distance between two lines in 3D.

two parameters s and t that yield these $\mathbf{q}_1$ and $\mathbf{q}_2$:

$$\mathbf{q}_1 = l_1(s) = \mathbf{p}_1 + s\mathbf{u}\,,$$
$$\mathbf{q}_2 = l_2(t) = \mathbf{p}_2 + t\mathbf{v}\,.$$

The vector $\mathbf{q}_1 - \mathbf{q}_2$ is perpendicular to $\mathbf{u}$ and $\mathbf{v}$:

$$\langle(\mathbf{p}_1 - \mathbf{p}_2) + s\mathbf{u} - t\mathbf{v},\ \mathbf{u}\rangle = 0\,,$$
$$\langle(\mathbf{p}_1 - \mathbf{p}_2) + s\mathbf{u} - t\mathbf{v},\ \mathbf{v}\rangle = 0\,,$$

leading to:

$$\langle\mathbf{p}_1 - \mathbf{p}_2, \mathbf{u}\rangle + s\|\mathbf{u}\|^2 - t\langle\mathbf{v}, \mathbf{u}\rangle = 0\,,$$
$$\langle\mathbf{p}_1 - \mathbf{p}_2, \mathbf{v}\rangle - t\|\mathbf{v}\|^2 + s\langle\mathbf{u}, \mathbf{v}\rangle = 0\,.$$

Hence, we have two equations and two unknowns. Therefore we solve for s and t:

$$\tilde{s} = \frac{\beta\langle\mathbf{w}, \mathbf{v}\rangle - \|\mathbf{v}\|^2\langle\mathbf{w}, \mathbf{u}\rangle}{\|\mathbf{u}\|^2\|\mathbf{v}\|^2 - \beta^2}\,,$$
$$\tilde{t} = \frac{\|\mathbf{u}\|^2\langle\mathbf{w}, \mathbf{v}\rangle - \beta\langle\mathbf{w}, \mathbf{u}\rangle}{\|\mathbf{u}\|^2\|\mathbf{v}\|^2 - \beta^2}\,,$$

where $\beta = \langle\mathbf{v}, \mathbf{u}\rangle$ and $\mathbf{w} = \mathbf{p}_1 - \mathbf{p}_2$. Finally, we have

$$\mathrm{dist}(l_1, l_2) = \|l_1(\tilde{s}) - l_2(\tilde{t})\|\,.$$

Cross product.[1] We end this chapter with a quick review of the notion of a cross product of two vectors, and present a related application of transforming planes. Formally, a cross product of vectors $\mathbf{a}$ and $\mathbf{b}$ is

$$\mathbf{a} \times \mathbf{b} = \|\mathbf{a}\|\|\mathbf{b}\| \sin\alpha\ \hat{\mathbf{n}}\,,$$

---

[1]Can be left out on the first reading.

where $\hat{\mathbf{n}}$ is a unit vector perpendicular to the plane containing $\mathbf{a}$ and $\mathbf{b}$, and $\sin \alpha$ is the sine of the smaller angle between $\mathbf{a}$ and $\mathbf{b}$ ($0 \leq \alpha \leq 2\pi$). It should be noted that, unlike the dot product, the cross product of two vectors is a vector. The vector $\mathbf{a} \times \mathbf{b}$ can be calculated by the determinant:

$$\mathbf{a} \times \mathbf{b} = \det \begin{bmatrix} \hat{\mathbf{i}} & \hat{\mathbf{j}} & \hat{\mathbf{k}} \\ a_1 & a_2 & a_3 \\ b_1 & b_2 & b_3 \end{bmatrix},$$

where $\hat{\mathbf{i}}, \hat{\mathbf{j}}$ and $\hat{\mathbf{k}}$ are unit vectors forming an orthogonal right-handed coordinate system, and $\mathbf{a} = a_1\hat{\mathbf{i}} + a_2\hat{\mathbf{j}} + a_3\hat{\mathbf{k}} = (a_1, a_2, a_3)$, and $\mathbf{b} = b_1\hat{\mathbf{i}} + b_2\hat{\mathbf{j}} + b_3\hat{\mathbf{k}} = (b_1, b_2, b_3)$.

An immediate application of the cross product is the definition of a plane by three non-collinear points $\mathbf{p}_1, \mathbf{p}_2$ and $\mathbf{p}_3$. A plane with an implicit equation $ax + by + cz + d = 0$ has a normal $\left(\frac{a}{d}, \frac{b}{d}, \frac{c}{d}\right)$. The direction of this normal vector is also equal to the cross product of $(\mathbf{p}_1 - \mathbf{p}_2) \times (\mathbf{p}_1 - \mathbf{p}_3)$, and the value of d can be determined by plugging the value of one of the points into the plane equation.

Now, given a transformation T and a plane $ax + by + cz + d = 0$, it is interesting to see how we can compute the equation of the transformed plane. A direct application of the above is to recompute the plane of $T(\mathbf{p}_1), T(\mathbf{p}_2)$ and $T(\mathbf{p}_3)$. However, there is a simpler method. Let us denote $\mathbf{a} = (a, b, c, d)$ and $\mathbf{x} = (x, y, z, 1)$, then we have $\mathbf{a}^\top \mathbf{x} = 0$. Similarly, the transformed plane is represented by $\mathbf{a}' = (a', b', c', d')$ and satisfies $\mathbf{a}'^\top T\mathbf{x} = 0$. Since we can write $\mathbf{a}^\top \mathbf{x} = 0$ as

$$\mathbf{a}^\top T^{-1} T\mathbf{x} = 0,$$

we get that

$$\mathbf{a}'^\top = \mathbf{a}^\top T^{-1},$$

or

$$\mathbf{a}' = T^{-\top}\mathbf{a}.$$

In other words, the coefficients of the transformed plane are merely the transposed inverse transform of the coefficients of the input plane.

Chapter 2

Linear Algebra?

Daniel Cohen-Or, Olga Sorkine-Hornung and Chen Greif

Our guess is that the term *linear algebra* means, for many of us, the name of a first-year undergraduate course, whose fundamental importance and benefit for solving geometric problems are not apparent to those taking it. In a linear algebra course, we are introduced to notions such as *vector space, orthogonal basis, subspaces, eigenvalues* and *eigenvectors*. Are these really useful?

Figure 2.1: Aligning the two shapes by minimizing the distance of corresponding landmark points.

Is there a close relation between linear algebra and geometry? The answers will probably be obvious after reading this book, but let's start from the following problem, illustrated in Figure 2.1: There are two shapes, one of a unicorn and one of a lion. The

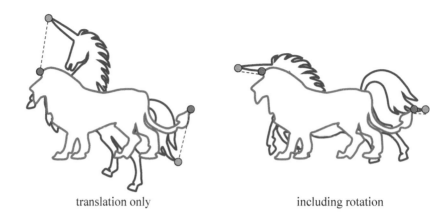

translation only   including rotation

Figure 2.2: Aligning the two shapes by a translation only (left). Aligning the shapes with a rotation as well (right). The transformation minimizes the sum of squared distances between corresponding points.

shapes are marked with two corresponding landmark-point pairs, illustrated by little red and blue circles. Corresponding points are connected by dotted lines. The problem now is: What is the best rigid transformation that aligns the landmarks of the two shapes? A rigid transformation can translate and rotate the shapes but not stretch or distort them whatsoever. Arguably the best transformation will minimize the length of the dotted lines, i.e., the mismatch error between corresponding points, or more precisely, the sum of the squared errors.

Let us regard the shapes as sets of points and try to align these sets using a linear transformation. By translating one shape so that its center is aligned with that of the other shape, we can get an alignment, as shown in Figure 2.2 (left). Subsequently, by rotating one shape, we get a better alignment, as shown in Figure 2.2 (right). This can be expressed as a least-squares problem and solved with various tools. The singular value decomposition (SVD), discussed in Chapter 4, is an important method for solving least-squares problems and other linear algebra problems. To understand what an SVD is, we need to understand what eigenvectors and eigenvalues are, which requires us to recall some fundamentals of linear algebra.

First, let us look at a related problem, illustrated in Figure 2.3. Given a set of points, find the line that best fits that set of points.

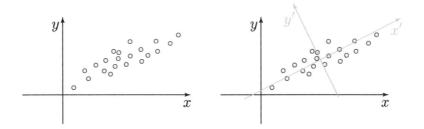

Figure 2.3: Left: A given set of points (left), and the line that best fits the set of points (right).

Again, more precisely, we want to minimize the sum of squared errors (see Figure 2.4). The error is the distance between a point and the nearest point on the line. This problem can be solved with principal component analysis (PCA). As we shall see in Chapter 4, PCA and SVD are tightly related.

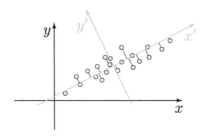

Figure 2.4: The line minimizes the squared distances from the line.

Fundamentals

First, let us informally define a *linear space* V:

- $V \neq \emptyset$: the space is a non-empty set of vectors;
- $\mathbf{v}, \mathbf{w} \in V \Rightarrow \mathbf{v} + \mathbf{w} \in V$: the space is closed under addition;
- $\mathbf{v} \in V$, α is scalar $\Rightarrow \alpha \mathbf{v} \in V$: the space is also closed under multiplication by scalars.

A formal definition includes axioms about associativity and distributivity of the sum and multiplication operators. It is very important to remember that $\mathbf{0} \in V$ always, which follows from the third condition above.

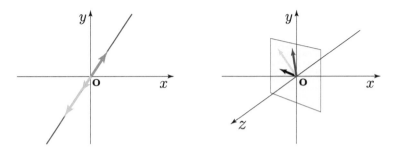

Figure 2.5: A line through the origin is a subspace of $\mathbb{R}^2$ (left). A plane through the origin is a subspace of $\mathbb{R}^3$ (right).

The xy plane is a simple example of a linear space. The notion of a *subspace* is also important and useful. For example, place a line l on the xy plane passing through the origin, $\mathbf{0}$. Then $L = \{\mathbf{p} - \mathbf{0} \mid \mathbf{p} \in l\}$, is a linear subspace of $\mathbb{R}^2$ (see Figure 2.5 (left)). A simple example of a subspace of $\mathbb{R}^3$ is a plane Π passing through the origin in 3D (see Figure 2.5 (right)). Note that in both cases, any addition of two vectors or their multiplication by a scalar results in a vector in the subspace. In particular it is true if the scalar is zero.

The vectors $\{\mathbf{v}_1, \mathbf{v}_2, \ldots, \mathbf{v}_k\}$ form a *linearly independent set* if $\alpha_1 \mathbf{v}_1 + \alpha_2 \mathbf{v}_2 + \cdots + \alpha_k \mathbf{v}_k = \mathbf{0} \iff \alpha_i = \mathbf{0} \; \forall i$. This means that none of the vectors can be obtained as a linear combination of the others. For example, parallel vectors are necessarily dependent since one can always be obtained by multiplying the other by a scalar.

Let $\{\mathbf{v}_1, \mathbf{v}_2, \ldots, \mathbf{v}_n\}$ be a linearly independent set. We say this set is a basis that *spans* the whole vector space

$$V = \{\alpha_1 \mathbf{v}_1 + \alpha_2 \mathbf{v}_2 + \cdots + \alpha_k \mathbf{v}_k\},$$

where the α_i are any scalars in $\mathbb{R}$. Any vector in V can be expressed as a unique linear combination of the basis. The number of basis vectors is called the *dimension* of V. Since any vector in V has a unique representation $\mathbf{v} = \alpha_1 \mathbf{v}_1 + \alpha_2 \mathbf{v}_2 + \cdots + \alpha_n \mathbf{v}_n$, the α_i are the *coordinates* or *coefficients* of the vector in that basis.

Two vectors are said to be *orthogonal* if their product is zero. A good basis (for computational purposes) often consists of orthogonal vectors. For example, the most popular choice of a basis for $\mathbb{R}^3$ consists of three unit-length orthogonal vectors $\mathbf{x}, \mathbf{y}, \mathbf{z}$, which are often denoted by $\hat{\mathbf{i}}, \hat{\mathbf{j}}, \hat{\mathbf{k}}$ or $\mathbf{e}_1, \mathbf{e}_2, \mathbf{e}_3$ (see Figure 2.6).

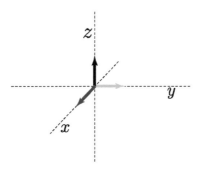

Figure 2.6: The standard basis in $\mathbb{R}^3$.

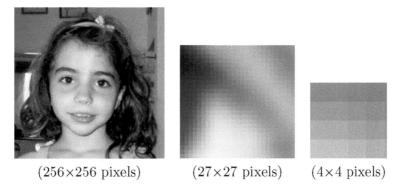

(256×256 pixels)   (27×27 pixels)   (4×4 pixels)

Figure 2.7: A digital image of size $N \times M$ is a vector in $\mathbb{R}^{NM}$.

Have you ever wondered what the advantage of having an orthogonal basis is? After all, any basis spans the space, and all its vectors are uniquely represented. This question will be discussed at the end of this chapter.

A vector space and a basis may have different forms. For example, the grayscale image consisting of N by M pixels (see Figure 2.7) is a point in $\mathbb{R}^{NM}$. The gray-level values at each pixel are the coefficients of its basis. A mini-image, or a 4×4 block in an image, has a 16-vector basis as illustrated in Figure 2.8 (top).

Unfolding the 4×4 vector basis reveals that, in fact, this is a standard basis. The bottom of Figure 2.8 shows a linear combination of the vectors composing the values of the mini-image. Figure 2.9 shows a different basis that consists of 64 vectors, each of which is a 8×8 image or a vector in $\mathbb{R}^{64}$. This basis, however, is non-standard but very popular. It is the *discrete cosine basis* used for JPEG encoding. Moreover, as we shall see in Chapter 5, this basis is *spectral*, allowing a frequency analysis.

The standard basis of 4×4 grayscale images:

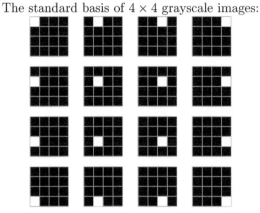

A linear combination (blend) of basis vectors compose an image:

Figure 2.8: Standard basis representation of grayscale images.

Linear spaces are associated with *inner products*. The inner (or dot) product between vectors $\mathbf{v}$ and $\mathbf{w}$ is denoted by $\langle \mathbf{v}, \mathbf{w} \rangle$, as we already saw in Chapter 1. If the coordinates of these vectors are given by $v_1, \cdots, v_n$ and $w_1, \cdots, w_n$, respectively, their inner product is:

$$\langle \mathbf{v}, \mathbf{w} \rangle = \sum_i v_i \, w_i \, .$$

As discussed in Chapter 1, the dot product of two vectors gives us useful information about angles and length, and generally about the metric of the linear space.

Matrices as linear operators

It is an understatement to say that linear operators are very important for geometric operations. Scaling, rotation and reflection are examples of simple linear operators. Formally, an operator $A : V \to W$ is a linear operator if $A(\mathbf{v} + \mathbf{w}) = A(\mathbf{v}) + A(\mathbf{w})$, and $A(\alpha \mathbf{v}) = \alpha A(\mathbf{v})$, which, in particular, implies that $A(\mathbf{0}) = \mathbf{0}$. Thus, translation is not a linear operator as it translates the origin. Figure 2.10 demonstrates that rotating the sum of two vectors is equivalent to the sum of the two rotated vectors.

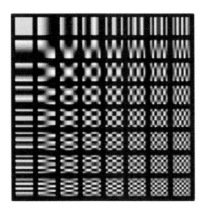

Figure 2.9: The discrete cosine transform (DCT) basis in $\mathbb{R}^{64}$.

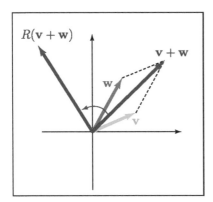 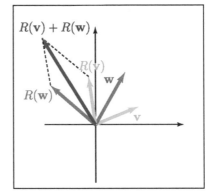

Figure 2.10: Rotation is a linear operator.

Consider the following vector column:

$$\mathbf{v} = \begin{bmatrix} \alpha_1 \\ \vdots \\ \alpha_n \end{bmatrix}.$$

We can write it as $\mathbf{v} = \alpha_1 \mathbf{v}_1 + \alpha_2 \mathbf{v}_2 + \cdots + \alpha_n \mathbf{v}_n$ with the standard basis vectors

$$\begin{bmatrix} 1 \\ 0 \\ \vdots \\ 0 \end{bmatrix}, \begin{bmatrix} 0 \\ 1 \\ \vdots \\ 0 \end{bmatrix}, \ldots, \begin{bmatrix} 0 \\ 0 \\ \vdots \\ 1 \end{bmatrix}.$$

Now, given an arbitrary vector $\mathbf{v} = \alpha_1 \mathbf{v}_1 + \alpha_2 \mathbf{v}_2 + \cdots + \alpha_n \mathbf{v}_n$, we

have
$$A(\mathbf{v}) = \alpha_1 A(\mathbf{v}_1) + \alpha_2 A(\mathbf{v}_2) + \cdots + \alpha_n A(\mathbf{v}_n),$$

which means that we can learn what the operator A does by look-ing at what it does to its basis vectors. This principle will be very useful in the following, and will now allow us to present the notion of a matrix to represent a linear operator A. Given

$$M_A = \begin{bmatrix} | & | & & | \\ A(\mathbf{v}_1) & A(\mathbf{v}_2) & \cdots & A(\mathbf{v}_n) \\ | & | & & | \end{bmatrix},$$

we have

$$\begin{bmatrix} | & | & & | \\ A(\mathbf{v}_1) & A(\mathbf{v}_2) & \cdots & A(\mathbf{v}_n) \\ | & | & & | \end{bmatrix} \begin{bmatrix} 1 \\ 0 \\ \vdots \\ 0 \end{bmatrix} = \begin{bmatrix} | \\ A(\mathbf{v}_1) \\ | \end{bmatrix}$$

and

$$\begin{bmatrix} | & | & & | \\ A(\mathbf{v}_1) & A(\mathbf{v}_2) & \cdots & A(\mathbf{v}_n) \\ | & | & & | \end{bmatrix} \begin{bmatrix} 0 \\ 1 \\ \vdots \\ 0 \end{bmatrix} = \begin{bmatrix} | \\ A(\mathbf{v}_2) \\ | \end{bmatrix},$$

and so on.

Matrix addition, subtraction and multiplication by a scalar are straightforward operations. Multiplication of a matrix by a col-umn vector can be presented as

$$A\mathbf{b} = \begin{bmatrix} a_{1,1} & \cdots & a_{1,n} \\ \vdots & \vdots & \vdots \\ a_{m,1} & \cdots & a_{m,n} \end{bmatrix} \begin{bmatrix} b_1 \\ \vdots \\ b_n \end{bmatrix} = \begin{bmatrix} \sum_i a_{1,i} b_i \\ \vdots \\ \sum_i a_{m,i} b_i \end{bmatrix} = \begin{bmatrix} \langle \text{row}_1, \mathbf{b} \rangle \\ \vdots \\ \langle \text{row}_m, \mathbf{b} \rangle \end{bmatrix},$$

but often it is better to look at $A\mathbf{b}$ as a linear combination of A's columns:

$$\begin{bmatrix} | & | & \cdots & | \\ \mathbf{a}_1 & \mathbf{a}_2 & \vdots & \mathbf{a}_n \\ | & | & \cdots & | \end{bmatrix} \begin{bmatrix} b_1 \\ \vdots \\ b_n \end{bmatrix} = b_1 \begin{bmatrix} | \\ \mathbf{a}_1 \\ | \end{bmatrix} + b_2 \begin{bmatrix} | \\ \mathbf{a}_2 \\ | \end{bmatrix} + \cdots + b_n \begin{bmatrix} | \\ \mathbf{a}_n \\ | \end{bmatrix}.$$

Transposition of A, denoted by $A^\top$, turns the rows into columns and vice versa, $a_{ij} \longmapsto a_{ji}$ (see Figure 2.11), and an important

$$\begin{bmatrix} a_{11} & \cdots & a_{1n} \\ \vdots & \ddots & \vdots \\ a_{m1} & \cdots & a_{mn} \end{bmatrix}^{\top} = \begin{bmatrix} a_{11} & \cdots & a_{m1} \\ \vdots & \ddots & \vdots \\ a_{1n} & \cdots & a_{mn} \end{bmatrix}$$

Figure 2.11: Transposing a matrix.

equality is $(AB)^{\top} = B^{\top}A^{\top}$. The transposition operation is useful for expressing an inner product between two vectors in a matrix form:

$$\langle \mathbf{v}, \mathbf{w} \rangle = \mathbf{v}^{\top}\mathbf{w} = \mathbf{w}^{\top}\mathbf{v}\,.$$

As we shall see in later chapters, we often solve problems by finding a vector $\mathbf{x}$ that satisfies a set of equations represented in a matrix form, $A\mathbf{x} = \mathbf{b}$. The system of equations $A\mathbf{x} = \mathbf{b}$ has a unique solution if the matrix A is non-singular (we assume A is a square matrix). An $n \times n$ matrix A is said to be non-singular if there exists a matrix B such that $AB = BA = I$, where I is the identity matrix. The matrix B is the inverse of A, denoted as A^{-1}. The determinant of a non-singular matrix satisfies $\det(A) \neq 0$. One important observation is that if A is non-singular, the rows of A are linearly independent, and so are the columns. This is an opportunity to remind the reader that the rank of a matrix is the number of linearly independent rows or columns (whichever is smaller).

When the matrix A is singular, there is no unique solution to the system: there are either infinitely many solutions or no solution. In the first case, the *null space* of the matrix features in the solution. A vector $\mathbf{z}$ is a null vector of A if $A\mathbf{z} = 0$. The null space is a set of linearly independent null vectors. For any $\mathbf{x}$ that satisfies $A\mathbf{x} = \mathbf{b}$, we have that $A(\mathbf{x} + \mathbf{z}) = \mathbf{b}$, hence the lack of uniqueness.

A matrix A is said to be an orthogonal matrix if $A^{-1} = A^{\top}$. It follows that for an orthogonal matrix A, $A^{\top}A = AA^{\top} = I$, and the columns of A are orthonormal vectors. Orthogonal matrices preserve the inner product, that is, lengths and angles:

$$\langle A\mathbf{v}, A\mathbf{w} \rangle = (A\mathbf{v})^{\top}(A\mathbf{w}) = \mathbf{v}^{\top}A^{\top}A\mathbf{w} = \mathbf{v}^{\top}I\mathbf{w} = \langle \mathbf{v}, \mathbf{w} \rangle\,.$$

Therefore, $\|A\mathbf{v}\| = \|\mathbf{v}\|$ and $\angle(A\mathbf{v}, A\mathbf{w}) = \angle(\mathbf{v}, \mathbf{w})$.

Any orthogonal 3×3 matrix represents a rotation around some axis (if $\det(A) = 1$), or a reflection (if $\det(A) = -1$). For example, a rotation by θ around the z-axis in $\mathbb{R}^3$ can be expressed in terms of a matrix as

$$R_\theta = \begin{bmatrix} \cos\theta & \sin\theta & 0 \\ -\sin\theta & \cos\theta & 0 \\ 0 & 0 & 1 \end{bmatrix},$$

which is an orthogonal matrix.

Norms

Vectors are arrays of numbers, but there are certain notions that are straightforward for numbers and less intuitively clear for vectors. One important issue is that of the magnitude or *norm* of a vector. The size or magnitude of a number is given by its absolute value; this is very intuitive. For vectors there are a few commonly used possibilities. The norm of a vector $\mathbf{u}$ is a scalar denoted by $\|\mathbf{u}\|$, which satisfies the following properties:

- $\|\mathbf{u}\| \geq 0$, and $\|\mathbf{u}\| = 0$ if and only if $\mathbf{u} = 0$, namely all of the components of $\mathbf{u}$ are zero.

- For any scalar α and a vector $\mathbf{u}$, $\|\alpha\mathbf{u}\| = |\alpha| \cdot \|\mathbf{u}\|$.

- The triangle inequality holds: $\|\mathbf{u} + \mathbf{v}\| \leq \|\mathbf{u}\| + \|\mathbf{v}\|$ for any two vectors $\mathbf{u}$ and $\mathbf{v}$.

Notice that for numbers, namely vectors of length one, the absolute value is trivially a valid norm that satisfies the above three properties.

For vectors, l_p-norms are commonly used and are defined as

$$\|\mathbf{u}\|_p = \left(\sum_{i=1}^{n} |u_i|^p \right)^{1/p}.$$

Commonly used norms are (see Figure 2.12):

- The *Euclidean norm*, which is also known as the l_2-*norm* or the *spectral norm*:

$$\|\mathbf{u}\|_2 = \sqrt{u_1^2 + u_2^2 + \cdots + u_n^2}.$$

Since this is the most commonly used norm, when no subscript appears next to the norm notation, it usually means the l_2-norm.

- The l_1-norm, which is the sum of absolute values:

$$\|\mathbf{u}\|_1 = |u_1| + |u_2| + \cdots + |u_n|;$$

- The l_∞-norm, which is also known as the *max norm* or *infinity norm*:

$$\|\mathbf{u}\|_\infty = \max_i |u_i|.$$

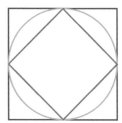

Figure 2.12: Unit circles for various norms.

Next, we briefly mention how *matrix norms* are defined. Especially important are induced matrix norms, which are defined as follows:

$$\|A\| = \max_{\mathbf{u} \neq 0} \frac{\|A\mathbf{u}\|}{\|\mathbf{u}\|},$$

where $\mathbf{u}$ is a vector. For example, the 2-norm of a matrix is defined as $\|A\|_2 = \max_{\mathbf{u} \neq 0} \frac{\|A\mathbf{u}\|_2}{\|\mathbf{u}\|_2}$, where $\|A\mathbf{u}\|_2$ and $\|\mathbf{u}\|_2$ are the previously defined Euclidean vector norms.

Eigenvectors and eigenvalues

We are now ready to discuss a couple of very important mathematical objects: eigenvectors and eigenvalues. Previously, we mentioned that any linear operator A can be analyzed by studying its operation on the basis vectors of the space. The eigenvectors of matrix A form a basis for the column space of A. Let A be a square $n \times n$ matrix. A vector $\mathbf{v}$ is an eigenvector of A if

$$A\mathbf{v} = \lambda\mathbf{v},$$

where λ is a scalar and $\mathbf{v}$ is a nonzero vector. The scalar λ is called an *eigenvalue* of $\mathbf{v}$. Notice that by this definition, eigenvectors are not unique; they are determined up to a multiplicative constant. It is common practice to normalize them, namely, set $\|\mathbf{v}\| = 1$.

An eigenvector spans an axis (subspace of dimension 1) that is invariant under A. It remains the same under the transformation.

Eigenvectors with the same eigenvalues form a linear subspace:

$$A\mathbf{v} = \lambda\mathbf{v}, \qquad A\mathbf{w} = \lambda\mathbf{w} \;\Rightarrow\; A(\mathbf{v} + \mathbf{w}) = \lambda(\mathbf{v} + \mathbf{w}).$$

How can we find eigenvalues? We look for λ for which there is a nontrivial solution to the equation $A\mathbf{x} = \lambda\mathbf{x}$. That is

$$A\mathbf{x} = \lambda\mathbf{x} \;\Leftrightarrow\; A\mathbf{x} - \lambda\mathbf{x} = \mathbf{0} \;\Leftrightarrow\; A\mathbf{x} - \lambda I\mathbf{x} = \mathbf{0} \;\Leftrightarrow\; (A - \lambda I)\mathbf{x} = \mathbf{0}.$$

So, a nontrivial solution exists if and only if $\det(A - \lambda I) = 0$. The expression $\det(A - \lambda I)$ can be developed into a polynomial of degree n called the *characteristic polynomial* of A. The roots of this characteristic polynomial are the eigenvalues of A. Therefore, there are always n eigenvalues (some or all of which may be complex). If n is odd, there is at least one real eigenvalue. Let us look at an example:

$$A = \begin{bmatrix} 1 & 0 & 2 \\ 3 & 0 & -3 \\ -1 & 1 & 4 \end{bmatrix}.$$

Then, $\det(A - \lambda I)$ is

$$\det(A) = \begin{bmatrix} 1 - \lambda & 0 & 2 \\ 3 & -\lambda & -3 \\ -1 & 1 & 4 - \lambda \end{bmatrix}$$

$$= (1 - \lambda)(-\lambda(4 - \lambda) + 3) + 2(3 - \lambda)$$

$$= (3 - \lambda)(\lambda^2 - 2\lambda + 3).$$

Note that the term $(\lambda^2 - 2\lambda + 3)$ cannot be factored over $\mathbb{R}$, but over $\mathbb{C}$ it does factor into $(1 + i\sqrt{2})(1 - i\sqrt{2})$, where $i = \sqrt{-1}$. So the three eigenvalues are $(3, 1 + i\sqrt{2}, 1 - i\sqrt{2})$.

Notice that even though the matrix is real, its eigenvalues may be complex. This often happens for non-symmetric matrices. On the other hand, symmetric matrices are guaranteed to have real eigenvalues and eigenvectors. Even though in an arbitrary case we will have complex eigenvalues, in geometric models the problem often leads to symmetric matrices. Therefore, it is useful to consider the special case of real eigenvalues and eigenvectors.

To compute the eigenvectors, we solve the equation $(A - \lambda I)\mathbf{x} = \mathbf{0}$ and get an eigenvector $\mathbf{x}$. Notice that $A - \lambda I$ is singular; we look for a *nonzero* $\mathbf{x}$. We also note that λ could be a multiple eigenvalue, in which case there could be a set of more than one linearly independent corresponding eigenvectors.

The set of all the eigenvalues of A, denoted by V, is called the *spectrum* of A.

Figure 2.13: $AV = VD$.

If A has n independent eigenvectors, then A is *diagonalizable*; see Figure 2.13. The set of equations $A\mathbf{v}_i = \lambda \mathbf{v}_i$ amounts to $AV = VD$, where D is a diagonal matrix, whose diagonal values are $\lambda_1, \lambda_2, \ldots, \lambda_n$. We can also write $A = VDV^{-1}$, and we see that A represents scalings along the eigenvector axes (see Figure 2.14).

Figure 2.14: $A = VDV^{-1}$.

To get a better intuitive feeling of the meaning of eigenvectors and the spectrum of A, let us look at a symmetric operator A. Since A is symmetric, all its eigenvalues are real and therefore we can easily visualize the operation. Figure 2.15 (left) illustrates the map of a circle under the A transform. In 2D it takes a circle to a rotated ellipse.

Looking at the mapping of an arbitrary vector (the black vector in Figure 2.15 (left)) does not tell us much about the transform. However, as illustrated in Figure 2.15 (right), the map of the two eigenvectors (the red and the green vectors) tells us a lot about

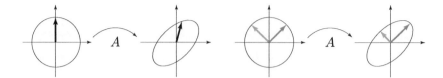

Figure 2.15: The map of a circle under a transform. An arbitrary vector does not tell much about the shape of the transform (left). The two eigenvectors correspond to the major/minor axes, and the eigenvalues give us the lengths of these axes (right).

the shape and the characteristics of the mapping. Moreover, as we shall see later on, the simple form of the spectra of A becomes very handy for various analyses.

Here are a few more useful facts. A is called *normal* if $AA^\top = A^\top A$. An important example is symmetric matrices $A = A^\top$. It can be proven that normal $n \times n$ matrices have exactly n linearly independent eigenvectors (over $\mathbb{C}$). If A is symmetric, all eigenvalues of A are real, and A has n linearly independent real orthogonal eigenvectors. Figure 2.14 illustrates this: A has orthonormal eigenvectors and thus V is an orthogonal matrix that transforms A into a diagonal matrix D.

A diagonalizable matrix is essentially a scaling. Often, matrices arising in applications are not diagonalizable; they do other things along with scaling (such as rotation). So, to understand how general matrices behave, mere eigenvalues are not enough. In Chapter 4 we will review the singular value decomposition (SVD), which will tell us how general linear transformations behave, among other things.

Finding the coordinates of a vector

Suppose we are given a basis $\mathbf{v}_1, \mathbf{v}_2, \mathbf{v}_3$ in $\mathbb{R}^3$. An arbitrary vector $\mathbf{a} \in \mathbb{R}^3$ has three coordinates denoted by, say, $\alpha_1, \alpha_2, \alpha_3$ and associated with the basis vectors:

$$\mathbf{a} = \sum_{i=1}^{n} \alpha_i \mathbf{v}_i .$$

To extract the α_i, we can use the linearity of the space and form inner products that give us

$$\langle \mathbf{a}, \mathbf{v}_j \rangle = \sum_{i=1}^{n} \alpha_i \langle \mathbf{v}_i, \mathbf{v}_j \rangle,$$

and thus we have

$$\langle \mathbf{a}, \mathbf{v}_1 \rangle = \alpha_1 \langle \mathbf{v}_1, \mathbf{v}_1 \rangle + \alpha_1 \langle \mathbf{v}_2, \mathbf{v}_1 \rangle + \alpha_3 \langle \mathbf{v}_3, \mathbf{v}_1 \rangle;$$
$$\langle \mathbf{a}, \mathbf{v}_2 \rangle = \alpha_2 \langle \mathbf{v}_1, \mathbf{v}_2 \rangle + \alpha_1 \langle \mathbf{v}_2, \mathbf{v}_2 \rangle + \alpha_3 \langle \mathbf{v}_3, \mathbf{v}_2 \rangle;$$
$$\langle \mathbf{a}, \mathbf{v}_3 \rangle = \alpha_3 \langle \mathbf{v}_1, \mathbf{v}_3 \rangle + \alpha_1 \langle \mathbf{v}_2, \mathbf{v}_3 \rangle + \alpha_3 \langle \mathbf{v}_3, \mathbf{v}_3 \rangle.$$

In other words, we have a linear system of three equations and three unknowns:

$$\begin{bmatrix} \langle \mathbf{v}_1, \mathbf{v}_1 \rangle & \langle \mathbf{v}_1, \mathbf{v}_2 \rangle & \langle \mathbf{v}_1, \mathbf{v}_3 \rangle \\ \langle \mathbf{v}_2, \mathbf{v}_1 \rangle & \langle \mathbf{v}_2, \mathbf{v}_2 \rangle & \langle \mathbf{v}_2, \mathbf{v}_3 \rangle \\ \langle \mathbf{v}_3, \mathbf{v}_1 \rangle & \langle \mathbf{v}_3, \mathbf{v}_2 \rangle & \langle \mathbf{v}_3, \mathbf{v}_3 \rangle \end{bmatrix} \begin{bmatrix} \alpha_1 \\ \alpha_2 \\ \alpha_3 \end{bmatrix} = \begin{bmatrix} \langle \mathbf{a}, \mathbf{v}_1 \rangle \\ \langle \mathbf{a}, \mathbf{v}_2 \rangle \\ \langle \mathbf{a}, \mathbf{v}_3 \rangle \end{bmatrix}.$$

Here is where orthogonality may come in handy. If the basis is orthogonal, then $\langle \mathbf{v}_i, \mathbf{v}_j \rangle = \begin{bmatrix} 1 & i=j \\ 0 & i \neq j \end{bmatrix}$, and we get the trivial identity matrix and all the above collapses into the simple form:

$$\begin{bmatrix} \alpha_1 \\ \alpha_2 \\ \alpha_3 \end{bmatrix} = \begin{bmatrix} \langle \mathbf{a}, \mathbf{v}_1 \rangle \\ \langle \mathbf{a}, \mathbf{v}_2 \rangle \\ \langle \mathbf{a}, \mathbf{v}_3 \rangle \end{bmatrix}.$$

The simplicity of this form is particularly attractive because no linear system needs to be solved (since the matrix to be inverted is the identity matrix). But orthogonality is important not only because it may simplify computations; it is also essential for numerical stability of computations, as it avoids an accumulation of roundoff errors.

Homogeneous coordinates

Earlier in this chapter, we claimed that a translation is not a linear operation. We attributed it to the fact that a translation operator T might translate the origin: $T(\mathbf{0}) \neq \mathbf{0}$. This is one of the reasons why *homogeneous coordinates* are commonly used in computer graphics. In homogeneous coordinates, translation is

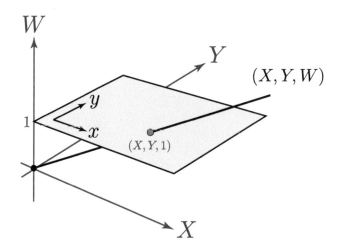

Figure 2.16: Homogeneous coordinates map from $\mathbb{R}^n$ to $\mathbb{R}^{n+1}$.

a linear operator! How does it work? First, let us define what homogeneous coordinates are.

Homogeneous coordinates map from $\mathbb{R}^n$ to $\mathbb{R}^{n+1}$. The best understood example is in 2D, shown in Figure 2.16. The 2D plane (x, y) is mapped to a parallel plane in 3D, (X, Y, W); for convenience, we choose $W = 1$. Instead of applying the transformations over $(x, y, 0)$, we apply them over (X, Y, W). Then, the transformed point is mapped back by

$$(X, Y, W) \mapsto ((X/W), (Y/W)).$$

The trick is that translating points in the plane $(X, Y, 1)$ never translates the origin of $\mathbb{R}^3$, and the operator that translates the vector $(x, y, 1)^\top$ by a and b to $(x + a, y + b, 1)^\top$ is a linear operator with a convenient matrix form:

$$\begin{bmatrix} 1 & 0 & a \\ 0 & 1 & b \\ 0 & 0 & 1 \end{bmatrix} \begin{bmatrix} x \\ y \\ 1 \end{bmatrix} = \begin{bmatrix} x + a \\ y + b \\ 1 \end{bmatrix}.$$

Note that the representation of a point in homogeneous coordinates is not unique. A point $(x, y, 1)$ is equivalent to (tx, ty, t) for any t. In other words, all the points along the line (tx, ty, t) (see Figure 2.16) are equivalent (or, if you'd like, homogeneous) under the projection back to the plane $W = 1$: $(x, y, 1) \equiv (tx, ty, t)$.

An interesting comment is that in a homogeneous coordinate system, all the affine transformations (i.e., translation, scaling, rotation, shear and their combinations) keep the points on the $W = 1$ plane.

Chapter 3

Least-Squares Solutions

Niloy J. Mitra

We start with a simple problem in geometry in the spirit of the ones discussed in Chapter 1. Given two lines in 2D, find their point of intersection (see Figure 3.1). Suppose the equations of the lines are

$$l_1 : a_1x + b_1y + c_1 = 0\,,$$
$$l_2 : a_2x + b_2y + c_2 = 0\,.$$

(3.1)

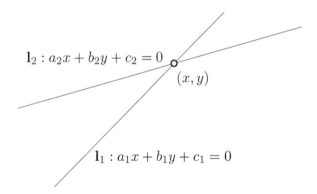

Figure 3.1: The point of intersection of two lines l_1 and l_2 lies on both lines and hence satisfies both equations. Given the two lines, we can solve for their intersection point using a simple linear system (see Equation (3.3)).

Stating that problem slightly differently, we are looking for a point (x, y) that lies on both lines, i.e., satisfies Equations (3.1).

31

We can express this in matrix form as

$$\begin{bmatrix} a_1 & b_1 \\ a_2 & b_2 \end{bmatrix} \begin{bmatrix} x \\ y \end{bmatrix} + \begin{bmatrix} c_1 \\ c_2 \end{bmatrix} = 0. \tag{3.2}$$

Using the notation $A = \begin{bmatrix} a_1 & b_1 \\ a_2 & b_2 \end{bmatrix}$ and $\mathbf{c} = \begin{bmatrix} c_1 \\ c_2 \end{bmatrix}$, we have

$$A \begin{bmatrix} x \\ y \end{bmatrix} + \mathbf{c} = 0 \quad \Longrightarrow \quad \begin{bmatrix} x \\ y \end{bmatrix} = -A^{-1}\mathbf{c}. \tag{3.3}$$

The desired solution is (x, y), the point where the two lines l_1 and l_2 intersect. If the matrix A is invertible, then indeed this solution is *unique*. However, when the two lines are parallel, the rows of matrix A become linearly dependent, leading to $\det(A) = 0$. The matrix degenerates, resulting in failure of the above method for finding the intersection point.

Let us look at a related problem: What happens when instead of two lines we have three or more lines? What does it mean? What if the three lines do not pass through a single point, but their points of (pairwise) intersection are close by (see Figure 3.2). We can try to look for a point that is *closest* to all the given lines. In such a scenario we often talk about a least-squares (LS) solution. This is an extremely powerful technique, and is applied in many diverse situations.

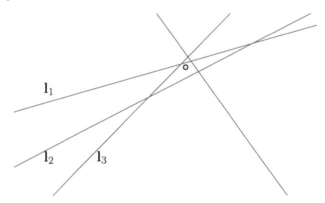

Figure 3.2: Given a set of lines, we are interested in finding the point that lies closest to all the lines. This point, obtained using an LS solution, is useful in many situations. Notice that the LS solution lies close to the average of the pairwise intersection points of the lines.

A more general problem is to find the best curve approximating a set of given points (see Figure 3.5). Instead of fitting a line, we now want to find the best curve from a family of them, i.e., optimize over a given family of curves. We will shortly see that again an LS solution can be employed to obtain such a robust fitting in the presence of measurement noise.

Let us discuss a similar scenario in a higher dimension: We are given a set of points sampled from a 3D surface (see Figure 3.3). However, for most geometry processing applications, we are interested in the underlying surface, and not just in any particular point sample of the surface. How can we model the underlying surface? One way is to locally fit a polynomial surface. Not surprisingly, once again LS fitting is the method of choice. Alternatively, one can try to solve for an exact fit surface—however, that is a bad idea because of noise in the data, and hence an approximate LS solution, which smooths out such measurement noise, is preferred.

The LS fitting method is simple and is based on elementary linear algebra described in the previous chapter. The procedure is surprisingly useful in many geometric modeling scenarios. Later in this chapter, we will further explore LS solutions for a range of apparently dissimilar problems.

Figure 3.3: Given a point set in 3D (left), we want to get a local approximation to the underlying sampled surface (right). This is achieved by locally fitting a polynomial surface using an LS fitting. Once such a polynomial fit is available, we can easily perform surface shading, upsampling, noise reduction, and other geometry processing tasks.

Line fitting

Let us start with a simple application of LS solutions: Given a set of n points $\mathbf{P} := \{\mathbf{p}_i\}$ on the plane, find the straight line that *best* approximates or fits the given data; see Figure 3.4 (left). Notice that this method can only give meaningful results if the underlying

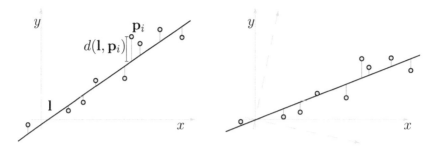

Figure 3.4: Given a set of points, we solve for the best-fitting line in the LS sense that minimizes the sum of squares of the *vertical distances* from the sample points (left). Notice that such an optimization function, being dependent on vertical distance, changes with a change in coordinate system (right). This is undesirable since the solution should depend only on the point set, and not on its physical embedding.

model from which data is being observed is indeed a line or close to being linear. We return to this issue later on. Even if this implicit assumption is valid, our current formulation is still incomplete since we have not described how to measure distance from a point to a line. Let us assume that the desired line is specified as $\mathbf{l} : y = ax + b$ (see Figure 3.4), then the vertical distance to the line from any point $\mathbf{p}_i = (x_i, y_i)$ is given by $d(\mathbf{l}, \mathbf{p}_i) := y_i - ax_i - b$. Our goal is to minimize the sum of squares of all such distances. Thus, our goal is to minimize

$$E(a, b) := \sum_{i=1}^{n} (y_i - ax_i - b)^2, \tag{3.4}$$

over all possible choices of a and b. Alternatively, we could have taken an expression of the form $\sum_i |d(\mathbf{l}, \mathbf{p}_i)|^k$ for some positive value k, and still get a meaningful result. However, most such choices of k result in a non-linear optimization. Later, we will learn a bit about non-linear optimization, but for now it is sufficient to note that non-linear optimizations are, in general, more challenging and harder than linear ones. Luckily, for $k = 2$ we get a *linear* system, and so we can use our linear algebra tools! Hence the name: least-*squares* solution. LS solutions owe their popularity to the fact that they can be solved using linear systems.

Before we try to minimize Equation (3.4), let us take a careful look at it. Do you see a problem with this formulation? Suppose the point sets are given with respect to another coordinate system. In that case, the objective function changes; see Figure 3.4 (right). Thus, in the current form, our solution depends on the choice of coordinate system. This should not be the case. The best-fitting line should only depend on the arrangement of the points, and remain unchanged with respect to the point set as long as its arrangement remains fixed. We will come back to this problem later on. For now, let us minimize Equation (3.4) to find the optimal choice of a and b. Recall from your calculus lessons that, at points where a function attains its extrema, maxima or minima, its partial derivatives ∂ (with respect to the function parameters) equal zero. In our setting, this amounts to

$$\partial E(a, b)/\partial a = \sum_{i=1}^{n} (-2x_i)(y_i - ax_i - b) = 0,$$
$$\partial E(a, b)/\partial b = \sum_{i=1}^{n} (-2)(y_i - ax_i - b) = 0. \qquad (3.5)$$

Simplifying the above equations and noting that a, b are the same for all the points, we get two linear equations in terms of the unknowns a and b:

$$\left(\sum_{i=1}^{n} x_i^2 \right) a + \left(\sum_{i=1}^{n} x_i \right) b = \sum_{i=1}^{n} x_i y_i,$$
$$\left(\sum_{i=1}^{n} x_i \right) a + nb = \sum_{i=1}^{n} y_i, \qquad (3.6)$$

which leads to a simple linear system of the form

$$\left(\sum_{i=1}^{n} \begin{bmatrix} x_i^2 & x_i \\ x_i & 1 \end{bmatrix} \right) \begin{bmatrix} a \\ b \end{bmatrix} = \sum_{i=1}^{n} \begin{bmatrix} x_i y_i \\ y_i \end{bmatrix}. \qquad (3.7)$$

Notice that in this equation $\{x_i\}$ and $\{y_i\}$ are known, and the matrices involving them can be built from the given point data set $\mathbf{P}$. The unknowns a, b are the solution of the linear system, and thus we have obtained the LS straight-line fit to the point set $\mathbf{P}$.

LS polynomial fits

We have studied how to extract an LS linear solution, i.e., given a point set, determine the *best*-fitting line in the LS sense. This is indeed a good estimate of the underlying model if the points are samples arising from a line segment. What if the points come from a parabola, or some other higher-order function (see Figure 3.5)? As we will shortly see, this approach is not all that different from what we have learned for the linear case.

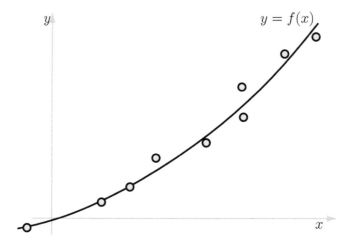

Figure 3.5: LS polynomial fit to a given set of points. The LS solution gives the best curve fit to a set of given points by optimizing over all curves from a family specified by a polynomial of the form $y = f(x)$.

Given a point set $\mathbf{P} := \{\mathbf{p}_i\}$, say we want to find the best-fitting m-th order polynomial function represented by $y = f(x) = \sum_{k=0}^{m} a_k x^k$. Again, taking vertical distances from points to the curve, we want to minimize the expression

$$E(a_0, \ldots, a_m) := \sum_{i=1}^{n} (f(x_i) - y_i)^2 = \sum_{i=1}^{n} \left(\sum_{k=0}^{m} a_k x_i^k - y_i \right)^2$$

over all choices of coefficients a_i. Using similar steps as before we get $(m + 1)$ equations by setting $\partial E(a_0, \ldots, a_m)/\partial a_k = 0$, for $k = 0, \ldots, m$. What might come as a surprise is that the solution is again a linear system! Let us understand the process using a simple example.

We will deal with a polynomial fit where $m = 2$, also referred to as a *parabolic fit*. The general polynomial of quadratic order can be written as $y = f(x) = \sum_{k=0}^{2} a_k x^k = a_0 + a_1 x + a_2 x^2$. By setting the partial derivatives to zero we get three equations, namely, $\partial \sum_i \left(y_i - \sum_{k=0}^{2} a_k x_i^k \right)^2 / \partial a_j$ for $j = 0, 1, 2$, respectively.

After simplification, the corresponding solution turns out to be

$$\left(\sum_{i=1}^{n}\begin{bmatrix} x_i^4 & x_i^3 & x_i^2 \\ x_i^3 & x_i^2 & x_i \\ x_i^2 & x_i & 1 \end{bmatrix}\right)\begin{bmatrix} a_2 \\ a_1 \\ a_0 \end{bmatrix} = \sum_{i=1}^{n}\begin{bmatrix} x_i^2 y_i \\ x_i y_i \\ y_i \end{bmatrix}. \qquad (3.8)$$

Now it is simple to get the corresponding LS solution by solving this linear system, where $\{x_i, y_i\}$ are known, while a_0, a_1, a_2 are unknowns. Note the similarity of the solution compared to the linear case given in Equation (3.7). However, unlike the case of a linear fit, for a polynomial fit, minimizing the sum of squares of the *normal* distance to the curve leads to a non-linear system. Hence, we use the vertical distance formulation, which also means that the solution depends on the coordinate system.

We continue with this approach and get an LS fit using any function $f(x)$. Suppose that the function is specified by parameters $a, b, c, \ldots$, such that the expression $f(x)$ *linearly* depends on the parameters. If this assumption is satisfied, then we can extract an LS solution using a linear system.

To repeat, we want to fit a function $f_{abc\ldots}(x)$ to data points $\{x_i, y_i\}$. Let us define $E(a, b, c, \ldots) := \sum_i (y_i - f_{abc\ldots}(x_i))^2$. Again, setting the partial derivatives to zero gives us

$$\partial E(a, b, c, \ldots)/\partial a = \sum_i (-2\partial f_{abc\ldots}(x_i)/\partial a)[y_i - f_{abc\ldots}(x_i)] = 0,$$

$$\partial E(a, b, c, \ldots)/\partial b = \sum_i (-2\partial f_{abc\ldots}(x_i)/\partial b)[y_i - f_{abc\ldots}(x_i)] = 0,$$

$$\vdots \qquad\qquad (3.9)$$

Since we assumed that $f(x)$ *linearly* depends on the parameters, terms like $\partial f(x_i)/\partial a$, $\partial f(x_i)/\partial b$, $\ldots$ work out such that the above system of equations is still linear in the parameters $a, b, \ldots$. For example, the function $f(x)$ can even be some complicated function like

$$f_{\lambda_1, \lambda_2, \lambda_3}(x) = \lambda_1 \sin^2 x + \lambda_2 \exp(-x^2/\sigma) + \lambda_3 x^{17},$$

or of the general form,

$$f_{\lambda_1, \lambda_2, \ldots, \lambda_k}(x) = \lambda_1 f_1(x) + \lambda_2 f_2(x) + \cdots + \lambda_k f_k(x).$$

Solving linear systems in the LS sense

Often in linear algebra we have to solve a set of m linear equations involving n unknowns. This can be represented compactly as $A\mathbf{x} = \mathbf{b}$, where A is a matrix of size $m \times n$, $\mathbf{b}$ is a vector of length m, and the n unknowns are represented by the vector $\mathbf{x}$. If $m = n$, and the matrix A has full rank, we know that the solution is $\mathbf{x} = A^{-1}\mathbf{b}$. What happens when $m > n$? What does it mean?

The above situation comes up in practice surprisingly often. When we have more equations than unknowns, we have an *over-determined system* of equations. Each of the equations typically contains some noise, or simply inaccurate observations. In such a situation we look for an LS solution, i.e., we want to solve the following optimization:

$$\min_{\mathbf{x}} \|A\mathbf{x} - \mathbf{b}\|^2 = \min_{\mathbf{x}}(\mathbf{x}^\top A^\top A\mathbf{x} - 2\mathbf{x}^\top A^\top \mathbf{b} + \mathbf{b}^\top \mathbf{b}). \qquad (3.10)$$

From Chapter 2 we know that $A\mathbf{x} = x_1\mathbf{a}_1 + \ldots + x_n\mathbf{a}_n$, where the $\mathbf{a}_i$ are the column vectors of A and the x_i are the components of the vector $\mathbf{x}$. This means that in Equation (3.10) we are looking for a vector in the subspace spanned by the columns of A that is *closest* to the vector $\mathbf{b}$, which is likely outside of the subspace. The smallest distance is attained if the difference vector $A\mathbf{x} - \mathbf{b}$ is *orthogonal* (or *normal*) to the column subspace, which means it should be orthogonal to all the spanning vectors, i.e., $\forall i, A\mathbf{x} - \mathbf{b} \perp \mathbf{a}_i$. These conditions are called the *normal equations*, and writing them down in matrix form is essentially equivalent to differentiating Equation (3.10) with respect to $\mathbf{x}$ using matrix differentiation rules:

$$\begin{aligned}
\forall i, \quad & \mathbf{a}_i^\top (A\mathbf{x} - \mathbf{b}) = 0 \\
\Rightarrow \quad & A^\top (A\mathbf{x} - \mathbf{b}) = 0 \\
\Rightarrow \quad & A^\top A\mathbf{x} = A^\top \mathbf{b}. \qquad (3.11)
\end{aligned}$$

The term $(A\mathbf{x} - \mathbf{b})$ is called the *residual vector*, the part of the data that is not explained by the LS solution. When $A^\top A$, which is of size $n \times n$, has full rank, the solution to the normal equations is obtained as $\mathbf{x} = (A^\top A)^{-1}A^\top \mathbf{b}$. If A has full rank, i.e., the columns of A are linearly independent, then $A^\top A$ also has full rank, and we can always take its inverse. Further, if we have different confidence or weights for the different equations, we can solve for the LS solution in the weighted sense. Next, we further explore this weighted approach.

The problem of outliers

LS fitting provides a robust solution in the presence of noise. However, often there are some isolated points in the given data set called *outliers*. An outlier is a data point that lies outside the overall pattern of the underlying distribution or data model. Suppose we are looking for a straight line best fitting a set of data points. Here we implicitly assume that the points do indeed come from a straight line, but allow for corruption due to noise. However, there can be isolated samples or points in the data that are actually not from the underlying model, a straight line in this case, that lie quite far away from it. Unfortunately, such a classification of good (inlier) and bad (outlier) points is not available to start with, and as a result the LS solution can be quite skewed; see Figure 3.6 (left). This happens because all points get equal treatment, i.e., both outliers and good measurements are considered equal, allowing spurious points to unduly influence the final solution.

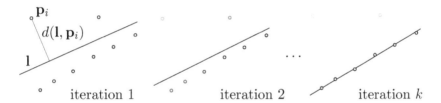

Figure 3.6: An LS line fit for a given set of unweighted points is not robust to outliers (left). As a solution, we progressively down-weigh points that are far from the fitted line, thus invoking a weighted LS solution. A monotonically decreasing function based on distance of points from the current LS line can be used as point weights for the next iteration. For example, one popular choice for the weight function is $w_{j+1} := \exp(-d(\mathbf{l}_j, \mathbf{p}_i)/\sigma)$, where j denotes the iteration number and σ is a user parameter controlling the falloff of the weights. Intuitively, points far from the current fit line are given less weight in the next iteration (indicated by fading gray colors in the figure). Thus, we simultaneously detect and down-weigh outliers while looking for the best solution in a weighted LS sense. This scheme can robustly handle data with a moderate amount of outliers, giving a good solution once the procedure converges (right).

A simple and natural way to handle this situation is to associate weights or confidence with points, i.e., we compute the LS line fit to a set of *weighted points*. At the beginning all points are considered equal, but as we know more about the points and their relation to the underlying model, we can progressively identify and down-weigh the outliers.

Let points $\mathbf{p}_i$ have weights $w_i \in [0, 1]$, then our new optimization becomes

$$\min_{\mathbf{l}} E(\mathbf{l}) := \sum_{i=1}^{n} w_i \, d(\mathbf{l}, \mathbf{p}_i)^2. \tag{3.12}$$

Using the vertical distance formulation, the new linear system formulation for fitting a straight line in the LS sense reduces to

$$\left(\sum_i \begin{bmatrix} w_i x_i^2 & w_i x_i \\ w_i x_i & w_i \end{bmatrix} \right) \begin{bmatrix} a \\ b \end{bmatrix} = \sum_i \begin{bmatrix} w_i x_i y_i \\ w_i y_i \end{bmatrix}. \tag{3.13}$$

When $w_i = 1$ for all points, this reduces to the unweighted case given by Equation (3.7).

If the input point set comes with corresponding confidence measures, we can directly apply the above method. Otherwise, we bootstrap the system with $w_i = 1$ for all points. For subsequent iterations, the distance of a point from the current solution is used as an indication of how likely this point is to be an outlier. A common choice of down-weighting scheme is $w_{j+1} = \exp(-d(\mathbf{l}_j, \mathbf{p}_i)/\sigma)$, where j denotes the iteration number and σ is a user parameter controlling the falloff of the weights. We stop iterating once the process converges. Even in the presence of a moderate number of outliers, this procedure typically behaves robustly; see Figure 3.6 (right). Notice that there are many different choices for weight functions that lead to similar behavior.

The idea of weighted LS is also applicable for a weighted least-squares solution. Recall that, corresponding to the n unknowns $\{x_i\}$ stacked together as vector $\mathbf{x}$, we had m equations. Now suppose each equation comes with a confidence value or weight, i.e., nonnegative value w_i. Thus we have m weights. The modified normal equation becomes

$$A^\top W A \mathbf{x} = A^\top W \mathbf{b}, \tag{3.14}$$

where W is an $m \times m$ diagonal matrix with the respective diagonal entries being the weights $\{w_i\}$.

Local surface fitting to 3D points

We have already seen how LS fitting is useful for fitting a polynomial to a set of points coming from a curve. Can we do something similar if the points come from a surface? More importantly, is it useful to locally fit a polynomial surface to a set of points in 3D? This scenario (see Figure 3.3) comes up often in computer graphics and digital shape acquisition, for example, when we get a rough sampling of a shape using a 3D laser or range scanner. Such a 3D point set $\mathbf{P} := \{\mathbf{p}_i\}$ can be treated as a sampling of the underlying shape or surface $\mathcal{S}$, similar to how a point set can be generated from an underlying curve. Given a point set or point cloud $\mathbf{P}$, if one can locally reconstruct or approximate the surface, then many geometric operations like shading, upsampling, intersection detection etc. become feasible. Interestingly, all such problems require us to locally estimate the surface normal at any point $\mathbf{p}$. Once we have a local surface normal, we have a local reference plane (see Figure 3.7 for an illustration of the process in 2D).

Figure 3.7: Illustration of local surface fitting with a 2D example. Given a set of points, for any point $\mathbf{p}$, we locally compute a linear LS fit (left). This gives us a local normal and tangent line at the point $\mathbf{p}$ (middle). Finally, we locally apply an LS polynomial fit using the computed normal tangent as the reference frame (right).

For any point $\mathbf{p} \in \mathbf{P}$, let $N_r(\mathbf{p}) = \{\mathbf{q}_i\}$ denote the points in $\mathbf{P}$ that are within a radius of r from $\mathbf{p}$, i.e., $\|\mathbf{p} - \mathbf{q_i}\| \leq r$. Let $\mathbf{n}$ be the unit normal to $\mathcal{S}$ at $\mathbf{p}$, and let the local tangent or the reference plane be denoted by $\mathbf{n}^\top \mathbf{x} + d = 0$. We can solve for this reference plane by looking for the plane that best fits the given set of point $N_r(\mathbf{p})$. Thus, our goal is to solve $\min_{\mathbf{n},d}(\mathbf{n}^\top \mathbf{q}_i + d)^2$ such that $\|\mathbf{n}\| = 1$. The solution will be discussed in detail in the next chapter; essentially, $\mathbf{n}$ is the smallest eigenvector of the covariance matrix $\sum_i (\mathbf{q}_i - \bar{\mathbf{q}})(\mathbf{q}_i - \bar{\mathbf{q}})^\top$, where $\bar{\mathbf{q}} = \sum_i \mathbf{q}_i/|N_r(\mathbf{p})|$. Note that the computations depend on the choice of r. If r is too small, the fit might be quite susceptible to noise, while if r is large, local variations such as sharp features can get averaged out.

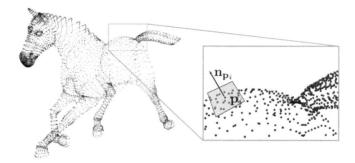

Figure 3.8: Using the LS method, for any point $\mathbf{p}_i$ in a given point set, we find the best-fitting local tangent plane whose normal is denoted by $\mathbf{n}_{\mathbf{p}_i}$. This tangent plane is subsequently used as a reference plane when locally computing a polynomial fit using LS fitting.

At any point, once we have a local reference plane, we are ready to locally fit a polynomial around the point. Using the reference plane as the xy plane, let the polynomial be given by $z = F(x, y) = ax^2 + bxy + cy^2 + dx + ey + f$, where $(a, \ldots, f)$ are the unknowns. Again, as in the case of polynomial fitting in 2D, given n observations $\{x_i, y_i\}$, the problem reduces to solving the following system of equations:

$$ax_1^2 + bx_1y_1 + cy_1^2 + dx_1 + ey_1 + f = z_1,$$
$$ax_2^2 + bx_2y_2 + cy_2^2 + dx_2 + ey_2 + f = z_2,$$
$$\ldots$$
$$ax_n^2 + bx_ny_n + cy_n^2 + dx_n + ey_n + f = z_n. \quad (3.15)$$

In the LS sense, we are trying to minimize $\sum_i \|z_i - F(x_i, y_i)\|^2$. By now we easily recognize the familiar LS setup.

Again, it is possible to have a procedure robust to outliers by introducing weights or confidence. In the weighted formulation, we minimize $\sum_i w_i \|z_i - F(x_i, y_i)\|^2$, where $w_i \geq 0$. Typically, we choose a weight function such that weights get smaller as the distance from the point $\mathbf{p}$ increases. A popular choice of such a weight function is exponential decay.

Concluding remarks

In this chapter, we have learned about Least-Squares fitting and solution. Recall that formulating the optimization using quadratic or *square* terms allows us to get a linear system after taking partial derivatives. The resultant linear system is simply solvable using basic linear algebra tools.

Advanced topic: Distance to lines[1]

Let us now address the remaining issue of how to make the LS line-fitting solution invariant under rigid transformations, i.e., make the formulation independent of the coordinate system. A general and detailed treatment of such problems will be presented in the next chapter using principal component analysis (PCA); here we give a short summary for the line-fitting case.

The solution is simple: For the distance of any point **p** to a line l, instead of taking the vertical distance, we take the *perpendicular* distance to l (see Figure 3.9).

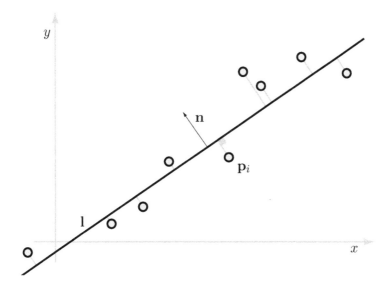

Figure 3.9: LS straight-line fitting to a given set of points using normal distance. The distance from any point $\mathbf{p}_i$ to line l is measured using the normal or the projected distance.

---

[1]This section makes use of Lagrange multipliers. You can skip this section in your first reading.

The line $\mathbf{l}$ consists of all points $\mathbf{x}$ such that $\mathbf{n}^\top\mathbf{x} + d = 0$, where $\mathbf{n}$ is a unit direction normal to the given line (see Figure 3.9). Our new optimization takes the form:

$$\min_{\mathbf{n},d} E(\mathbf{n}, d) = \sum_{i=1}^{n} (\mathbf{n}^\top\mathbf{p}_i + d)^2, \quad \text{s.t. } \|\mathbf{n}\| = 1. \tag{3.16}$$

Similar to our previous approach, we use the minimization conditions: $\partial E(\mathbf{n}, d)/\partial\mathbf{n} = 0$ and $\partial E(\mathbf{n}, d)/\partial d = 0$. Let us first investigate the second condition:

$$\partial E(\mathbf{n}, d)/\partial d = 0$$
$$\Rightarrow \quad 2\sum_i (\mathbf{n}^\top\mathbf{p}_i + d) = 0$$
$$\Rightarrow \quad nd = -\mathbf{n}^\top\sum_i \mathbf{p}_i$$
$$\Rightarrow \quad d = -\mathbf{n}^\top\bar{\mathbf{p}}, \tag{3.17}$$

where $\bar{\mathbf{p}} = \sum_i \mathbf{p}_i/n$ is the centroid or the mean of the given point set $\mathbf{P}$.

Combining Equations (3.16) and (3.17), we get a simpler optimization:

$$\min_{\mathbf{n}} E(\mathbf{n}) = \sum_{i=1}^{n} (\mathbf{n}^\top\mathbf{p}_i - \mathbf{n}^\top\bar{\mathbf{p}})^2 = \sum_{i=1}^{n} (\mathbf{n}^\top\tilde{\mathbf{p}}_i)^2, \quad \text{s.t. } \|\mathbf{n}\| = 1, \tag{3.18}$$

where $\tilde{\mathbf{p}}_i = \mathbf{p}_i - \bar{\mathbf{p}}$ denote the mean centered points. Observe that now we are only optimizing over a choice of unit vector $\mathbf{n}$.

Using an optimization trick where we penalize a solution not satisfying the side condition of unit norm ($\|\mathbf{n}\| = 1$), we reformulate the problem as

$$\min_{\mathbf{n}} \left(\sum_i (\mathbf{n}^\top\tilde{\mathbf{p}}_i)^2 + \lambda(1 - \mathbf{n}^\top\mathbf{n}) \right) = \min_{\mathbf{n}}(\mathbf{n}^\top C\mathbf{n} + \lambda(1 - \mathbf{n}^\top\mathbf{n})), \tag{3.19}$$

λ being the Lagrange multiplier, and $C = \sum_i \tilde{\mathbf{p}}_i\tilde{\mathbf{p}}_i^\top$ is the covariance matrix of the given point set.

Setting the partial derivative of the above expression with respect to $\mathbf{n}$ to zero, and using rules of matrix differentiation, we arrive at

$$2C\mathbf{n} - 2\lambda\mathbf{n} = 0 \quad \Rightarrow \quad C\mathbf{n} = \lambda\mathbf{n}. \tag{3.20}$$

Thus $\mathbf{n}$ is one of the eigenvectors of the covariance matrix C (see Chapter 4). Note that substituting Equation (3.20) in Equation (3.19), the objective function reduces to simply λ. Since our goal is to minimize the objective function, we select the *smallest eigenvector*, i.e., the eigenvector corresponding to the smallest eigenvalue. Once we have $\mathbf{n}$, we obtain d as $d = -\mathbf{n}^\top \bar{\mathbf{p}}$.

Chapter 4

PCA and SVD

Olga Sorkine-Hornung

In this chapter, we introduce two related tools from linear algebra that have become true workhorses in countless areas of scientific computing: principal component analysis (PCA) and singular value decomposition (SVD). Essentially, we will talk about a decomposition of a given matrix into several factors that are easy to analyze and reveal important properties of the matrix and hence the data, or the problem in which the matrix arises. Surprisingly, the singular value decomposition is often excluded from undergraduate linear algebra curricula, even though it is so widely used not only in geometric modeling and computer graphics, but also in computer vision, image processing, machine learning and many other applications.

Principal Component Analysis

PCA uses the spectral decomposition that we described earlier in Chapter 2 to discover the linear structure of a given data set. To phrase it in another way, given a set of (possibly multi-dimensional) data points, PCA finds a good orthogonal basis that represents these points well, or is well aligned with the data. PCA discovers dominant linear directions, along which the data has strong variation. Let us look at several simple examples where such an orthogonal basis is useful, before defining the process formally.

Recall the problem from the previous chapter (Chapter 3): Given a set of points in the plane, how do we find a straight line that passes as close as possible to these points (and thus approximates the data set well)? As you can see in Figure 4.1(a), PCA gives the answer by finding an orthogonal coordinate system in the plane, such that the first axis, x', is the best approximating line for the points. By *best* here we mean smallest sum of square distances between the points and the line (the distances are measured by orthogonal projection), as opposed to just vertical distance. Similarly, one can use PCA to discover the best approximating plane for 3D points (see Figure 4.1(b)).

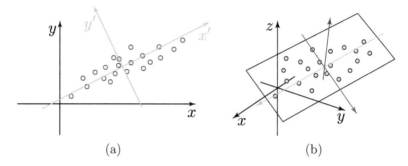

(a)   (b)

Figure 4.1: Principal component analysis finds orthogonal axes that represent the input points well in terms of linear structures. Specifically, (a) the first principal component (the x' axis) is the best least-square-distance approximation of the data points by a straight line, and (b) the first two principal components form a plane that best approximates the input points.

The first immediate application of PCA in graphics is the computation of oriented bounding boxes for 3D objects. A bounding box is a very simple and crude approximation of an object's shape: while the object may have a complex representation with thousands of polygons, its bounding box only has six planar faces. Bounding boxes are useful for fast collision detection and any other case where the rough (and conservative) dimensions of the object are needed. If the bounding boxes of two shapes do not intersect, then the two shapes cannot intersect either; the advantage here is that the intersection test against a box is very quick compared to testing the actual shape. On the other hand, the bounding box should be as tight as possible to prevent "false alarms" where the bounding boxes intersect but the actual objects do not. Computing the bounding box with respect to arbitrary coordinate axes may provide a poor fit, whereas PCA provides us a good set of axes for this task (see Figure 4.2).

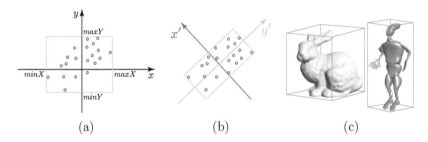

Figure 4.2: (a) An axis-aligned bounding box can be easily computed, but if the coordinate system is arbitrary, the box might not be very tight. (b) PCA gives us better axes, along which a tighter bounding box can be computed. (c) Bounding boxes are useful to roughly represent 3D objects for fast collision detection, approximate distance computations and more.

PCA has many further uses in graphics and geometric modeling, such as local frame estimation for point-cloud data, compact representation of animated geometric sequences, finding intuitive tuning parameters in facial modeling and animation, and more. We will look into some of these applications after taking a quick dive into the mathematical definition of PCA and the intuition behind it.

Let us denote our data points by $\mathbf{x}_1, \mathbf{x}_2, \ldots, \mathbf{x}_n \in \mathbb{R}^d$. Each $\mathbf{x}_i$ is a d-dimensional column vector; d could be 2 or 3 if we are looking at points sampled from some geometric object, but our data could also be high-dimensional: for example, each point could represent an entire shape, with all its sample points concatenated into one long vector.

PCA defines a new orthogonal coordinate system for our data. First of all, we define the origin of this coordinate system to be the center of mass of all the data points:

$$\mathbf{m} = \frac{1}{n} \sum_{i=1}^{n} \mathbf{x}_i \, .$$

The new origin $\mathbf{m}$ is the best zero-order approximation for our data, in the sense that it is the point that has the minimal sum of square distances to all the data points:

$$\mathbf{m} = \operatorname*{argmin}_{\mathbf{x}} \sum_{i=1}^{n} \|\mathbf{x}_i - \mathbf{x}\|^2.$$

Next, let us find the *directions* that represent our data best. For this, the scatter (or covariance) matrix $S \in \mathbb{R}^{d \times d}$ is defined. Denote $\mathbf{y}_i = \mathbf{x}_i - \mathbf{m}$ (these are the vectors from the new origin to our data points); the scatter matrix is

$$S = YY^{\top},$$

where Y is a $d \times n$ matrix that has the vectors $\mathbf{y}_i$ as its columns:

$$Y = \begin{bmatrix} | & | & & | \\ \mathbf{y}_1 & \mathbf{y}_2 & \cdots & \mathbf{y}_n \\ | & | & & | \end{bmatrix}.$$

Although it is not immediately obvious, in a way, the matrix S measures the variance, or the scatter of the points $\mathbf{x}_i$ in different directions. To understand why, let us look at an arbitrary line ℓ that passes through the center of mass $\mathbf{m}$, and project our data points onto this line. The variance of the projected points $\mathbf{x}_i'$ measures how far they are spread away from the center:

$$\mathrm{var}(\ell) = \frac{1}{n} \sum_{i=1}^{n} \|\mathbf{x}_i' - \mathbf{m}\|^2.$$

As explained in Figure 4.3, lines ℓ that align well with the data will have higher variance of projected points, and the minimal scatter will naturally happen in an orthogonal direction.

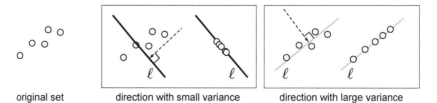

original set   direction with small variance   direction with large variance

Figure 4.3: The scatter matrix S measures the variance of the data set along any direction ℓ. We define *variance along a direction* as the variance (or the amount of scatter) among the points when projected onto that direction.

Now let us look more closely into the expression of variance. Define the direction of line ℓ by $\mathbf{v}$ and assume that $\mathbf{v}$ is normalized, i.e., $\|\mathbf{v}\| = 1$. The parametric line representation for ℓ would then be $\ell(t) = \mathbf{m} + t\mathbf{v}$, with the parameter t varying over all the real numbers. Therefore, we can replace the projected points $\mathbf{x}_i'$ in the expression of variance as follows:

$$\|\mathbf{x}_i' - \mathbf{m}\| = \left| \frac{\langle \mathbf{v}, \mathbf{x}_i - \mathbf{m} \rangle}{\|\mathbf{v}\|} \right| = |\langle \mathbf{v}, \mathbf{y}_i \rangle| = |\mathbf{v}^\top \mathbf{y}_i| .$$

Recall that we have learned about this relationship between orthogonal projection and inner product in Chapter 1. So now we can look at the variance $\text{var}(\ell)$ and discover the scatter matrix S in it:

$$\text{var}(\ell) = \frac{1}{n} \sum_{i=1}^{n} \|\mathbf{x}_i' - \mathbf{m}\|^2 = \frac{1}{n} \sum_{i=1}^{n} (\mathbf{v}^\top \mathbf{y}_i)^2 = \frac{1}{n} \|\mathbf{v}^\top Y\|^2$$

$$= \frac{1}{n} (\mathbf{v}^\top Y)(\mathbf{v}^\top Y)^\top = \frac{1}{n} \mathbf{v}^\top Y Y^\top \mathbf{v} = \frac{1}{n} \langle S\mathbf{v}, \mathbf{v} \rangle .$$

So we learn that the quadratic form defined by S in fact measures the scatter of the data points projected onto a line defined by a normalized direction vector. Perhaps the only non-obvious transition in the algebraic manipulations above was $\sum_{i=1}^{n}(\mathbf{v}^\top \mathbf{y}_i)^2 = \|\mathbf{v}^\top Y\|^2$. To see this, recall that the $\mathbf{y}_i$s are the columns of the matrix Y, so $\mathbf{v}^\top Y$ is an n-dimensional row vector whose entries are $\mathbf{v}^\top \mathbf{y}_i$. Therefore, the square length of this row vector is precisely the sum of squared norms of $\mathbf{v}^\top \mathbf{y}_i$.

Now that we know that

$$\text{var}(\ell) = \langle S\mathbf{v}, \mathbf{v} \rangle ,$$

we can find the directions of maximal and minimal scatter in the data by checking where the quadratic form attains its extrema. Here, we have an important theorem to help us.

Theorem: Let $f : \{\mathbf{v} \in \mathbb{R}^d, \text{ s.t. } \|\mathbf{v}\| = 1\} \to \mathbb{R}$ be a real-valued function on the unit sphere in $\mathbb{R}^d$, defined by $f(\mathbf{v}) = \langle S\mathbf{v}, \mathbf{v} \rangle$, where S is a real, symmetric $d \times d$ matrix. Then, the extrema of f are attained at the (normalized) eigenvectors of S.

For the sake of brevity, the proof of this theorem is omitted.[1] What is important here is that as a consequence, we learn a number of things: the directions of maximal/minimal scatter in our data points are precisely the eigenvectors of the scatter matrix S; because S is symmetric, it has an orthogonal eigenbasis, so the

---

[1] The theorem can be easily proven using Lagrange multipliers.

directions of extremal variance are pairwise orthogonal and form an orthonormal basis for the space $\mathbb{R}^d$. Let us name the eigenvalues of S, $\lambda_1, \lambda_2, \ldots, \lambda_d$, and the eigenvectors $\mathbf{v}_1, \mathbf{v}_2, \ldots, \mathbf{v}_d$. The extremal values of the variance are, therefore,

$$\langle S\mathbf{v}_i, \mathbf{v}_i \rangle = \langle \lambda_i \mathbf{v}_i, \mathbf{v}_i \rangle = \lambda_i \langle \mathbf{v}_i, \mathbf{v}_i \rangle = \lambda_i \, .$$

This means that we can sort the extremal directions $\mathbf{v}_i$ in the order of importance simply by sorting them according to the corresponding eigenvalues. In particular, the direction of maximal variance corresponds to the eigenvector with the largest eigenvalue, and the direction of minimal variance corresponds to the eigenvector with the smallest eigenvalue.

To summarize, here is the procedure for finding the principal components of a given data set $\mathbf{x}_1, \ldots, \mathbf{x}_n \in \mathbb{R}^d$:

1. Compute the center of mass of the data points,

$$\mathbf{m} = \frac{1}{n} \sum_{i=1}^{n} \mathbf{x}_i.$$

2. Translate all the points so that the origin is at $\mathbf{m}$:

$$\mathbf{y}_i = \mathbf{x}_i - \mathbf{m}, \; i = 1, \ldots, n.$$

3. Construct the $d \times d$ scatter matrix $S = YY^\top$, where Y is the matrix whose columns are the $\mathbf{y}_i$.
4. Compute the spectral decomposition: $S = V\Lambda V^\top$.
5. Sort the eigenvalues in decreasing order: $\lambda_1 \geq \lambda_2 \geq \cdots \geq \lambda_d$.
6. The corresponding orthonormal eigenvectors, $\mathbf{v}_1, \ldots, \mathbf{v}_d$, are the principal components, sorted in the order of significance (quality of alignment with the data).

Now let us talk about the meaning of the eigenvalues and their distribution. Analyzing the relationship between the eigenvalues helps us discover something about the structure of the data set. First, all the eigenvalues are nonnegative, because the scatter matrix S is positive semi-definite (it is a "square" of the matrix Y). If some eigenvalues are close to zero, this means that the data has no strong components in the subspace spanned by the corresponding eigenvectors. For example, if our data points are in 3D but actually are all sampled from a plane, there will be one very small

Figure 4.4: There is no obvious linear direction in the data (left). There is a clear direction in the data (right).

eigenvalue (theoretically, it should be zero if there is no noise in the data), and the corresponding eigenvector will be orthogonal to that plane. In general, if all the eigenvalues are similar to each other, it means that there is no preferred direction in the data, or no underlying linear structures. On the other hand, if some eigenvalues are large compared to others, it means that the data could be well described by a linear subspace (a line, a plane and so on): even though the data comes from $\mathbb{R}^d$, it has in fact intrinsically lower dimensionality, or is close to a lower-dimensional subspace. Figure 4.4 illustrates both cases in 2D. The decomposition of scatter matrix S of the points is

$$S = V \begin{bmatrix} \lambda_1 & \\ & \lambda_2 \end{bmatrix} V^\top.$$

On the left, there is no obvious linear direction in the data, and so $\lambda_1 \approx \lambda_2$. On the right there is a clear direction in the data, and the decomposition of scatter matrix S of the points is

$$S = V \begin{bmatrix} \lambda & \\ & \mu \end{bmatrix} V^\top,$$

where μ is close to zero.

Applications. Let us briefly discuss two useful applications. First, an application presented at the beginning of this chapter: finding oriented bounding boxes. We simply compute the bounding box with respect to the axes defined by the eigenvectors, and the origin is at the center of mass, as we discussed previously (see Figure 4.5).

Another very useful application is to reduce the dimensionality of the data, as illustrated in Figure 4.6. Projecting the data points onto the subspace spanned by the leading eigenvectors is a good

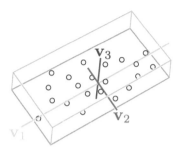

Figure 4.5: A bounding box can be placed around a set of points by aligning it with the eigenvectors of the scatter matrix of the points.

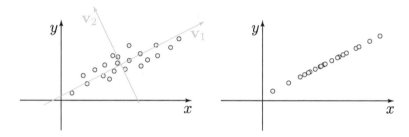

Figure 4.6: The projected data set (right) approximates the original data set (left) in a lower-dimensional subspace.

approximation of the original data set by a set embedded in a lower dimension.

In general dimension d, the eigenvalues are sorted in descending order:

$$\lambda_1 \geq \lambda_2 \geq \cdots \geq \lambda_d .$$

The eigenvectors are sorted accordingly. To get an approximation of dimension $d' \leq d$, we take the d' first eigenvectors and look at the subspace they span ($d' = 1$ is a line, $d' = 2$ is a plane). To get an approximating set, we project the original data points onto the chosen subspace: Let

$$\mathbf{x}_i = \mathbf{m} + \alpha_1 \mathbf{v}_1 + \alpha_2 \mathbf{v}_2 + \cdots + \alpha_{d'} \mathbf{v}_{d'} + \cdots + \alpha_d \mathbf{v}_d .$$

Then the $\mathbf{x}_i$ are projected by zeroing out the small eigenvalues:

$$\mathbf{x}_i = \mathbf{m} + \alpha_1 \mathbf{v}_1 + \alpha_2 \mathbf{v}_2 + \cdots + \alpha_{d'} \mathbf{v}_{d'} + 0 \cdot \mathbf{v}_{d'+1} + \cdots + 0 \cdot \mathbf{v}_d .$$

Much more on dimensionality reduction is discussed in Chapter 9.

Another application of PCA is approximating normals of point clouds. 3D scanners provide raw point-cloud data. To assign a

Figure 4.7: To assign a normal to a point, we can compute the PCA of its local neighboring points.

normal to a point, we can compute the PCA of its local neighboring points. The direction of the normal is then simply the direction of the third eigenvector, associated with the smallest eigenvalue, as illustrated in Figure 4.7. Note that additional inside–outside analysis needs to be done to properly orient the normal so that it points outside of the shape.

Singular Value Decomposition

Now we are ready to understand what singular value decomposition (SVD) is. In the following, we will first develop some basic intuition about it, then define it more formally, and finally show some applications.

Given a transformation A, we want to know what it does or analyze it geometrically. For that we need some simple and comprehensible representation of the matrix that corresponds to the transformation A. Let us observe what A does to some vectors. Since $A(\alpha\mathbf{v}) = \alpha A(\mathbf{v})$, it is enough to look at vectors $\mathbf{v}$ of unit length. A (non-singular) linear transformation always takes hyper-spheres to hyper-ellipses (depicted in Figure 4.8). Thus, one good way to understand what A does is to find which vectors are mapped to the main axes of the ellipsoid.

If we are lucky, then $A = V\Lambda V^\top$, V being orthogonal, meaning that A is symmetric, and the eigenvectors of A are the axes of the ellipse (as shown in Figure 4.9). In this case, A is just a scaling matrix. The eigendecomposition of A tells us which orthogonal

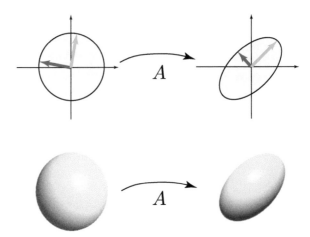

Figure 4.8: A linear transformation always takes hyper-spheres to hyper-ellipses.

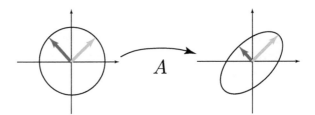

Figure 4.9: If the transformation is symmetric, its eigenvectors are the axes of the ellipse.

axes it scales, and by how much:

$$A = (\mathbf{v}_1 \ \mathbf{v}_2 \ \cdots \ \mathbf{v}_n) \begin{bmatrix} \lambda_1 & & & \\ & \lambda_2 & & \\ & & \ddots & \\ & & & \lambda_n \end{bmatrix} (\mathbf{v}_1 \ \mathbf{v}_2 \ \cdots \ \mathbf{v}_n)^\top$$

or in a compact notation: $A\mathbf{v}_i = \lambda_i \mathbf{v}_i$.

However, in general, A also contains rotations, not just scaling along orthogonal axes, and the decomposition is more involved (see Figure 4.10). The decomposition of a general transformation

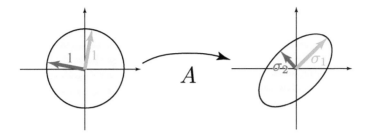

Figure 4.10: In general, a transformation may contain rotations and scales.

A then looks like this:

$$A = (\mathbf{u}_1 \; \mathbf{u}_2 \; \cdots \; \mathbf{u}_n) \begin{bmatrix} \sigma_1 & & & \\ & \sigma_2 & & \\ & & \ddots & \\ & & & \sigma_n \end{bmatrix} (\mathbf{v}_1 \; \mathbf{v}_2 \; \cdots \; \mathbf{v}_n)^\top,$$

i.e., $A = U\Sigma V^\top$, where Σ is a diagonal matrix with values σ_i that are real and nonnegative, and U and V are orthogonal matrices. This is the singular value decomposition (SVD). The nice thing is that SVD exists for any matrix of any dimension. In Figure 4.11 we show the SVD for a square matrix.

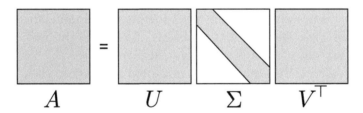

Figure 4.11: Singular value decomposition of a square matrix.

The diagonal values of Σ, $\sigma_1, \sigma_2, \ldots, \sigma_n$ are called the *singular values*. It is customary to sort them: $\sigma_1 \geq \sigma_2 \geq \ldots \geq \sigma_n \geq 0$.

For rectangular matrices, we have two forms of SVD. The reduced SVD has a rectangular U matrix with orthonormal columns (as shown in Figure 4.12). The reduced SVD is a cheaper form for computation and storage.

The full SVD has its U matrix completed to a full square orthogonal matrix, and we pad the Σ matrix by zero rows accordingly (see Figure 4.13).

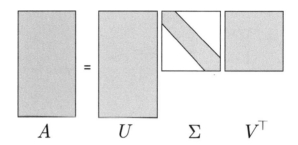

Figure 4.12: A reduced SVD.

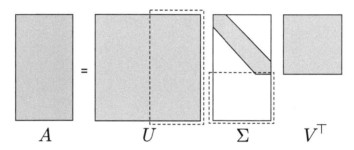

Figure 4.13: A full SVD.

Some applications of SVD. As we said previously, SVD is the "workhorse" of linear algebra. Once you have SVD, you immediately have many things: matrix inverse, and hence solution to square linear systems; you can tell the numerical rank of a matrix; you can solve least-squares problems. SVD directly provides the necessary data for PCA analysis, and much more.

Let $A = U \Sigma V^\top$ be a square non-singular matrix, then the inverse A^{-1} is

$$\left(U \Sigma V^\top \right)^{-1} = \left(V^\top \right)^{-1} \Sigma^{-1} U^{-1}$$

$$= V \begin{bmatrix} \frac{1}{\sigma_1} & & \\ & \ddots & \\ & & \frac{1}{\sigma_n} \end{bmatrix} U^\top.$$

Then to solve the linear system of equations $A\mathbf{x} = \mathbf{b}$, we have $\mathbf{x} = V\Sigma^{-1}U^\top\mathbf{b}$.

The rank of A is the number of nonzero singular values. If there are very small singular values, then A is close to being singular. We can set a threshold t, so that $\text{numeric\_rank}(A) = \#\{\sigma_i | \sigma_i > t\}$.

If the rank of A is smaller than n, then A is singular and it maps the entire space $\mathbb{R}^n$ onto some subspace, like a plane (so A is some sort of a projection).

Earlier we discussed computing the principal directions of a set of points $\mathbf{x}_i$, or in fact the vectors $\mathbf{y}_i$ (which are $\mathbf{x}_i$ shifted by the center of mass $\mathbf{m}$): $\mathbf{y}_i = \mathbf{x}_i - \mathbf{m}$. Let Y be the matrix whose columns are the vectors $\mathbf{y}_i$:

$$
Y = \begin{bmatrix} | & | & & | \\ \mathbf{y}_1 & \mathbf{y}_2 & \cdots & \mathbf{y}_n \\ | & | & & | \end{bmatrix}.
$$

Suppose we have the SVD of Y: $Y = U\Sigma V^\top$. Then the scatter matrix $S = YY^\top$ simplifies to

$$
\begin{aligned}
YY^\top &= U\Sigma V^\top (U\Sigma V^\top)^\top \\
&= U\Sigma V^\top V\Sigma^\top U^\top \\
&= U(\Sigma\Sigma^\top)U^\top.
\end{aligned}
$$

Thus, the column vectors of U are the principal components of the data set, and commonly they are sorted by the size of the singular values of Y.

Now, we can go back to the question raised in Chapter 2 and show how SVD can solve it: Given two objects with corresponding landmarks (shown in Figure 4.14 (left)), how can we find a rigid transformation that aligns them (Figure 4.14 (right))? When the objects are aligned, the lengths of the line segments connecting the landmarks are small. Therefore, we can solve a least-squares problem. Let $\mathbf{p}_i$ and $\mathbf{q}_i$ be the corresponding sets of points. We seek a translation vector $\mathbf{t}$ and a rotation matrix R so that $\sum_{i=1}^{n} \|\mathbf{p}_i - (R\mathbf{q}_i + \mathbf{t})\|^2$ is minimized.

It turns out that we can solve for the translation and rotation separately. If $(R, \mathbf{t})$ is the optimal transformation, then the points $\{\mathbf{p}_i\}$ and $\{R\mathbf{q}_i + \mathbf{t}\}$ have the same centers of mass. To see that, let $\mathbf{p} = \frac{1}{n}\sum_{i=1}^{n} \mathbf{p}_i$ and $\mathbf{q} = \frac{1}{n}\sum_{i=1}^{n} \mathbf{q}_i$. Given optimal R and $\mathbf{t}$,

$$
\mathbf{p} = \frac{1}{n}\sum_{i=1}^{n}(R\mathbf{q}_i + \mathbf{t}) = R\left(\frac{1}{n}\sum_{i=1}^{n}\mathbf{q}_i\right) + \mathbf{t} = R\mathbf{q} + \mathbf{t},
$$

and $\mathbf{t} = \mathbf{p} - R\mathbf{q}$. The same can be shown by differentiating our least-squares objective with respect to $\mathbf{t}$.

Figure 4.14: Aligning two shapes in correspondence by a rigid transformation.

To find R, let us assume that the corresponding points have been shifted so that their centers of mass align. Now we want to find R that minimizes

$$\sum_{i=1}^{n} \|\mathbf{p}_i - R\mathbf{q}_i\|^2.$$

Let

$$H = \sum_{i=1}^{n} \mathbf{q}_i \mathbf{p}_i^\top.$$

Given the SVD of $H = U\Sigma V^\top$, the optimal orthogonal transformation is $R = VU^\top$. Below we see why.

Given the orthogonality of R, we have $R^\top R = I$ and hence

$$\sum_{i=1}^{n} \|\mathbf{p}_i - R\,\mathbf{q}_i\|^2 = \sum_{i=1}^{n} (\mathbf{p}_i - R\,\mathbf{q}_i)^\top (\mathbf{p}_i - R\,\mathbf{q}_i)$$

$$= \sum_{i=1}^{n} \mathbf{p}_i^\top \mathbf{p}_i - \mathbf{p}_i^\top R\,\mathbf{q}_i - \mathbf{q}_i^\top R\,\mathbf{p}_i + \mathbf{q}_i^\top R^\top R\,\mathbf{q}_i$$

$$= \sum_{i=1}^{n} \mathbf{p}_i^\top \mathbf{p}_i - \mathbf{p}_i^\top R\,\mathbf{q}_i - \mathbf{q}_i^\top R\,\mathbf{p}_i + \mathbf{q}_i^\top \mathbf{q}_i\,.$$

The first and last terms, $\mathbf{p}_i^\top \mathbf{p}_i$ and $\mathbf{q}_i^\top \mathbf{q}_i$, do not depend on R, so we can ignore them in the minimization of $\sum_{i=1}^{n} \|\mathbf{p}_i - R\mathbf{q}_i\|^2$. Thus, the minimization reduces to

$$\min_{R} \sum_{i=1}^{n} (-\mathbf{p}_i^\top R\,\mathbf{q}_i - \mathbf{q}_i^\top R\,\mathbf{p}_i) = \max_{R} \sum_{i=1}^{n} (\mathbf{p}_i^\top R\,\mathbf{q}_i + \mathbf{q}_i^\top R\,\mathbf{p}_i)\,.$$

Since the second term $\mathbf{q}_i^\top R \mathbf{p}_i$ is a scalar, we have

$$\mathbf{p}_i^\top R \mathbf{q}_i = (\mathbf{p}_i^\top R \mathbf{q}_i)^\top = \mathbf{p}_i^\top R \mathbf{q}_i \, ,$$

which implies that

$$\operatorname*{argmax}_R \sum_{i=1}^n 2\mathbf{p}_i^\top R \mathbf{q}_i = \operatorname*{argmax}_R \sum_{i=1}^n \mathbf{p}_i^\top R \mathbf{q}_i \, .$$

Simplifying further,

$$\sum_{i=1}^n \mathbf{p}_i^\top R \mathbf{q}_i = \operatorname{Trace}\left(\sum_{i=1}^n R \mathbf{q}_i \mathbf{p}_i^\top\right) = \operatorname{Trace}\left(R \sum_{i=1}^n \mathbf{q}_i \mathbf{p}_i^\top\right) \, ,$$

where $\operatorname{Trace}(A) = \sum_{i=1}^n A_{ii}$.

Hence, we want to find R that maximizes $\operatorname{Trace}(RH)$. It is known that if M is symmetric positive definite (all eigenvalues of M are positive) and B is any orthogonal matrix, then

$$\operatorname{Trace}(M) \geq \operatorname{Trace}(BM) \, .$$

Thus, let us find R so that RH is symmetric positive definite. Then we know for sure that $\operatorname{Trace}(RH)$ is maximal. If $H = U\Sigma V^\top$ is the SVD, we define $R = VU^\top$. Now let us check RH:

$$RH = (VU^\top)(U\Sigma V^\top) = V\Sigma V^\top,$$

which is a symmetric matrix and its eigenvalues are positive, meaning that RH is symmetric positive definite. If you want to go deeper into the algebra of this problem, refer to our technical note "Least-Squares Rigid Motion Using SVD", http://igl.ethz.ch/projects/ARAP/svd_rot.pdf

Practical computation of principal components. Computing the SVD is expensive, and we always need to pay attention to the dimensions of the matrix. In many applications we need to compute the principal components for very large matrices, and that can be extremely computationally expensive or even infeasible. Suppose, for example, that each vector represents the pixels of an image: then the vector length is approximately $16K$ for a rather small-size image of 128×128 pixels, and the scatter matrix is a huge matrix of dimensions $16K \times 16K$. However, the rank of the scatter matrix is much smaller, as it is bounded by the number of vectors that participate in the analysis, say $N = 100$, and the principal components of the N vectors can be computed with much less effort. Suppose our $16K \times 100$ matrix is denoted by A. Instead of computing the PCA of the huge scatter matrix of $S = AA^\top$, we can compute the PCA of $A^\top A$, which is much smaller: a matrix of $N \times N$ elements (100×100 in our example instead of $16K \times 16K$). Now, let us explain how.

Denote by $\mathbf{v}_i$ the eigenvectors of the huge matrix $AA^\top$, and by $\mathbf{u}_i$ the eigenvectors of the small matrix $A^\top A$.

We have $A^\top A\mathbf{u}_i = \lambda_i \mathbf{u}_i$. Multiplying both sides of the equation by A, we get

$$AA^\top(A\mathbf{u}_i) = A\lambda_i \mathbf{u}_i = \lambda_i(A\mathbf{u}_i).$$

Thus, $\mathbf{v}_i = A\mathbf{u}_i$ are the eigenvectors of $AA^\top$. This means that we can compute the eigenvectors of $A^\top A$ and multiply them by A to get the eigenvectors of $AA^\top$.

Chapter 5

Spectral Transform

Hao (Richard) Zhang

In the past several decades, we have witnessed an explosion of multimedia content, such as music, images, and video. One of the main driving forces behind this phenomenon is the availability of effective *compression* algorithms for such data. While conventional text compression such as ZIP exploits predictabilities in the bit representation of the input text, the compression of visual data, such as an image, typically focuses on organizing regularities in its visual perception.

Typically, high compression ratios are achieved when certain data loss can be tolerated. One of the most popular approaches to lossy compression relies on *transform-based* techniques, where data given in the time or spatial domain is transformed into the spectral or frequency domain, and data compression can be achieved by properly lowering the accuracy of the representations in the frequency domain. A prominent example in image compression is JPEG, where high-frequency content in an image is represented with lower accuracy, since the human eye is less sensitive to it. Although in practice, JPEG image compression involves such steps as color space transformation and downsampling (to accentuate brightness over chrominance), block splitting (dividing an image into small blocks) and quantization (rounding real numbers to integers), the main ingredient of the approach is the use of the discrete cosine transform (DCT) [Jai89] to transform an image into the frequency domain. In Figure 5.1, we illustrate the effect of eliminating (i.e., zeroing) particular percentages of the high-frequency DCT coefficients, achieving the effect of compression.

Original   80%   90%   95%   98%   99%

Figure 5.1: The Lena image can be compressed by eliminating the indicated percentages of DCT coefficients. The images shown are the results of reconstruction using the remaining DCT coefficients.

As we can see from Figure 5.1, the compressed images appear to be blurred or smoothed versions of the original with certain sharp features and high-frequency details (e.g., the hat accessories) removed. Indeed, each compressed image is constructed as a weighted sum of a set of *basis images*. The basis images represent intensity oscillations at varying frequencies, as shown in Figure 5.2,

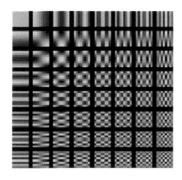

Figure 5.2: Plots of 64 DCT bases for 8×8 grayscale images.

where we plot the 64 DCT basis images for 8×8 grayscale images. For a natural image, such as the Lena in Figure 5.1, the most significant transform coefficients, which are the weights in the weighted sum, correspond to low-frequency basis images. On the other hand, those that characterize high-frequency contents can typically be eliminated without introducing noticeable quality degradation in the reconstructed image. In general, when high-frequency contents are removed, the image is smoothed.

As we can see from Figure 5.1, with only the first 20% of the (low-frequency) basis images used, corresponding to a compression ratio of 5:1, we already obtain a faithful reconstruction of the original image. With the other steps taken by conventional JPEG compression and adding entropy coding to the quantized transform coefficients, much higher compression rates can be obtained.

With recent advances in 3D data acquisition technology, highly detailed geometric models can be obtained via laser scanning. Efficient storage and delivery of these large geometric models motivate the development of geometry compression methods. Similar to image compression, visual perception plays a key role in geometry compression, where intensity variations over an image are replaced by geometric undulations over the surface of a geometric model. It is then natural to ask whether transform-based lossy compression techniques, such as a JPEG-like scheme, can be applied to geometry. This question is answered in this chapter, where we present tools for spectral transform of 3D geometric data.

Before dealing with surfaces of 3D shapes, we first consider curves representing 2D shapes. This allows us to reduce the 2D shape analysis problem to the study of 1D functions specifying the contour. We introduce the notion of Laplacian smoothing of these contours and show how spectral analysis based on 1D discrete Laplacian operators can perform smoothing as well as compression. The relationship between spectral analysis based on Laplacians and the classical Fourier transform is then revealed. Spectral transforms derived from eigendecomposition of mesh Laplacian operators are then defined, and we discuss their applications to mesh compression and filtering.

Spectral analysis in 1D

Consider the seahorse shape depicted by a closed contour, shown in Figure 5.3 (left). The contour is represented as a sequence of 2D points (the contour vertices) that are connected by straight line segments (the contour segments), as illustrated by the zoomed-in view. Now suppose that we wish to remove the rough features over the shape of the seahorse and obtain a smoothed version, such as the one shown in Figure 5.3 (right).

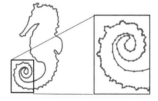

Figure 5.3: A seahorse shape with rough features (left) and a smoothed version (right).

Laplacian smoothing. A simple procedure to accomplish this is to repeatedly connect the midpoints of successive contour seg-

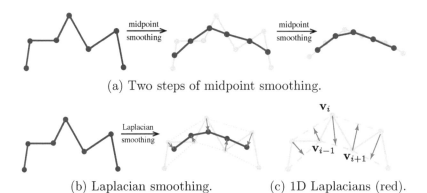

(a) Two steps of midpoint smoothing.

(b) Laplacian smoothing.   (c) 1D Laplacians (red).

Figure 5.4: (a) Two steps of midpoint smoothing are equivalent to (b) one step of Laplacian smoothing. (c) The 1D discrete Laplacian vectors are shown in red.

ments; we refer to this as *midpoint smoothing*. Figure 5.4(a) illustrates two such steps.

After two steps of midpoint smoothing, each contour vertex $\mathbf{v}_i$ is moved to the midpoint of the line segment connecting the midpoints of the original contour segments adjacent to $\mathbf{v}_i$. Specifically, let $\mathbf{v}_{i-1} = [x_{i-1}, y_{i-1}]$, $\mathbf{v}_i = [x_i, y_i]$, and $\mathbf{v}_{i+1} = [x_{i+1}, y_{i+1}]$ be three consecutive contour vertices. Then the new vertex $\mathbf{v}_i'$ after two steps of midpoint smoothing is given by a *local averaging*,

$$\mathbf{v}_i' = \frac{1}{2}\Big(\frac{1}{2}\left(\mathbf{v}_{i-1} + \mathbf{v}_i\right)\Big) + \frac{1}{2}\Big(\frac{1}{2}\left(\mathbf{v}_i + \mathbf{v}_{i+1}\right)\Big) = \frac{1}{4}\mathbf{v}_{i-1} + \frac{1}{2}\mathbf{v}_i + \frac{1}{4}\mathbf{v}_{i+1}.$$
(5.1)

The vector between $\mathbf{v}_i$ and the midpoint of the line segment connecting the two vertices adjacent to $\mathbf{v}_i$, shown as a red arrow in Figure 5.4(c), is called the *1D discrete Laplacian* at $\mathbf{v}_i$:

$$\delta(\mathbf{v}_i) = \frac{1}{2}(\mathbf{v}_{i-1} + \mathbf{v}_{i+1}) - \mathbf{v}_i.$$
(5.2)

As we can see, after two steps of midpoint smoothing, each contour vertex is displaced by half of its 1D discrete Laplacian, as shown in Figures 5.4(b)–(c); this is referred to as *Laplacian smoothing*. The smoothed version of the seahorse in Figure 5.3 was obtained by applying 10 steps of Laplacian smoothing.

Some natural questions one would ask include: Why is Laplacian smoothing removing the rough features (or noise) in the data? What happens if we apply Laplacian smoothing repeatedly and in the limit? To answer these questions, we will first need to express

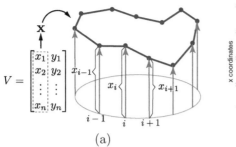
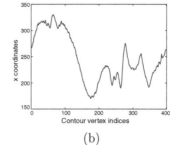

(a)   (b)

Figure 5.5: The x component of the contour coordinate matrix V, $\mathbf{x}$, can be viewed as (a) a 1D periodic signal defined over uniform samples along a circle. (b) Here is a conventional 1D plot for the seahorse contour from Figure 5.3.

our problem and algorithms algebraically and then apply analysis using the algebraic tools we reviewed in previous chapters.

1D signal and spectral transform. Let us denote the contour vertices by a coordinate matrix V, which has n rows and two columns, where n is the number of contour vertices and the two columns correspond to the x and y coordinates of the vertices. Let us denote by x_i (respectively, y_i) the x (respectively, y) coordinate of a vertex $\mathbf{v}_i$, $i = 1, \ldots, n$. For analysis purposes, let us only consider the x component of V, denoted by $\mathbf{x}$; the treatment of the y component is similar.

We treat the vector $\mathbf{x}$ as a discrete 1D signal. Since the contour is closed, we can view $\mathbf{x}$ as a periodic 1D signal defined over uniformly spaced samples along a circle; see Figure 5.5(a). We sort the contour vertices in counterclockwise order and plot $\mathbf{x}$ as a conventional 1D signal, designating an arbitrary element in $\mathbf{x}$ to start the indexing. Figure 5.5(b) shows such a plot for the x coordinates of the seahorse shape ($n = 401$) from Figure 5.3.

We can now express the 1D Laplacians (Equation (5.2)) for all the vertices using an $n \times n$ matrix L, called the *1D discrete Laplacian operator*:

$$\delta(\mathbf{x}) = L\mathbf{x} = \begin{bmatrix} 1 & -\frac{1}{2} & 0 & \cdots & \cdots & 0 & -\frac{1}{2} \\ -\frac{1}{2} & 1 & -\frac{1}{2} & 0 & \cdots & \cdots & 0 \\ \vdots & \vdots & \vdots & \vdots & \vdots & \vdots & \vdots \\ 0 & \cdots & \cdots & 0 & -\frac{1}{2} & 1 & -\frac{1}{2} \\ -\frac{1}{2} & 0 & \cdots & \cdots & 0 & -\frac{1}{2} & 1 \end{bmatrix} \mathbf{x}. \qquad (5.3)$$

Laplacian smoothing can then be expressed as applying a smoothing operator S to the signal $\mathbf{x}$, resulting in a new contour represented by $\mathbf{x}' = S\mathbf{x}$. The smoothing operator

$$
S = \begin{bmatrix}
\frac{1}{2} & \frac{1}{4} & 0 & \cdots & \cdots & 0 & \frac{1}{4} \\
\frac{1}{4} & \frac{1}{2} & \frac{1}{4} & 0 & \cdots & \cdots & 0 \\
\vdots & \vdots & \vdots & \vdots & \vdots & \vdots & \vdots \\
0 & \cdots & \cdots & 0 & \frac{1}{4} & \frac{1}{2} & \frac{1}{4} \\
\frac{1}{4} & 0 & \cdots & \cdots & 0 & \frac{1}{4} & \frac{1}{2}
\end{bmatrix} \tag{5.4}
$$

is related to the Laplacian operator by $S = I - \frac{1}{2}L$.

To analyze the behavior of Laplacian smoothing, in particular what happens in the limit, we utilize the set of basis vectors formed by the eigenvectors of L. This leads to a framework for *spectral analysis* of geometry. From linear algebra, we know that since L is symmetric, it has real eigenvalues and a set of real and orthogonal eigenvectors that form a basis. Any vector of size n can be expressed as a linear sum of these basis vectors. We are particularly interested in such an expression for the coordinate vector $\mathbf{x}$. Denote by $\mathbf{e}_1, \mathbf{e}_2, \ldots, \mathbf{e}_n$ the normalized eigenvectors of L, corresponding to eigenvalues $\lambda_1, \lambda_2, \ldots, \lambda_n$, and let E be the matrix whose columns are the eigenvectors. Then we can express $\mathbf{x}$ as a linear combination of the eigenvectors,

$$
\mathbf{x} = \sum_{i=1}^{n} \tilde{x}_i\, \mathbf{e}_i = \begin{bmatrix} | & | & & | \\ \mathbf{e}_1 & \mathbf{e}_2 & \cdots & \mathbf{e}_n \\ | & | & & | \end{bmatrix} \begin{bmatrix} \tilde{x}_1 \\ \tilde{x}_2 \\ \vdots \\ \tilde{x}_n \end{bmatrix} = E\,\tilde{\mathbf{x}}. \tag{5.5}
$$

Since the eigenvectors form an orthonormal basis, E is invertible and $E^{-1} = E^{\top}$. Hence, we can write

$$
\tilde{\mathbf{x}} = E^{\top}\mathbf{x}.
$$

This represents a transform from the original signal $\mathbf{x}$ to a new signal $\tilde{\mathbf{x}}$ in terms of a new basis, the basis given by the eigenvectors of L. We call this a *spectral transform*. For each i, the *spectral transform coefficient* is

$$
\tilde{x}_i = \mathbf{e}_i^{\top}\mathbf{x}. \tag{5.6}
$$

That is, the spectral coefficient $\tilde{x}_i$ is obtained as a *projection* of the signal $\mathbf{x}$ along the direction of the i-th eigenvector $\mathbf{e}_i$. In Figure 5.6, we plot the first 8 eigenvectors of L, sorted by increasing

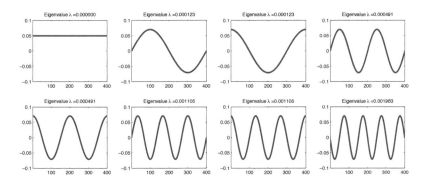

Figure 5.6: Plots of the first 8 eigenvectors of the 1D discrete Laplacian operator ($n = 401$) given in Equation (5.3). They are sorted by eigenvalue λ, shown above each plot.

eigenvalues, where $n = 401$, matching the size of the seahorse shape from Figure 5.3. The indexing of elements in each eigenvector follows the same contour vertex indexing as $\mathbf{x}$, the coordinate vector, as plotted in Figure 5.5(b) for the seahorse. It is worth noting that aside from an agreement on indexing, the Laplacian operator L and the eigenbasis vectors do not depend on $\mathbf{x}$, which specifies the geometry of the contour. L, as defined in Equation (5.3), is completely determined by n, the size of the input contour, and a vertex ordering.

As we can see, the eigenvector corresponding to the zero eigenvalue is a constant vector. As the eigenvalues increase, the eigenvectors start oscillating as sinusoidal curves at higher and higher frequencies. Note that the eigenvalues of L repeat (multiplicity 2) after the first one, hence the corresponding eigenvectors of these repeated eigenvalues are not unique. One particular choice of the eigenvectors reveals a connection of our spectral analysis to the classical Fourier analysis; this will be discussed later on.

Signal reconstruction and compression. With the spectral transform of a coordinate signal defined in Equation (5.5), we can now look at filtering and compression of a 2D shape represented by a contour. Analogous to JPEG image compression, we can obtain compact representations of a contour by retaining the leading (low-frequency) spectral transform coefficients and eliminating the rest. Given a signal $\mathbf{x}$ as in Equation (5.5), the signal reconstructed by using the k leading coefficients is

$$\mathbf{x}^{(k)} = \sum_{i=1}^{k} \tilde{x}_i \, \mathbf{e}_i, \quad k < n \,. \tag{5.7}$$

This represents a compression of the contour geometry as only $k < n$ coefficients need to be stored to approximate the original shape. We can quantify the information loss by measuring the l_2 error or Euclidean distance between $\mathbf{x}$ and $\mathbf{x}^{(k)}$,

$$\|\mathbf{x} - \mathbf{x}^{(k)}\| = \left\| \sum_{i=k+1}^{n} \tilde{x}_i \, \mathbf{e}_i \right\| = \sqrt{\sum_{i=k+1}^{n} \tilde{x}_i^2} \,.$$

The last equality is easy to obtain if we note the orthogonality of the eigenvectors, i.e., $\mathbf{e}_i^\top \mathbf{e}_j = 0$ whenever $i \neq j$. Also, since the $\mathbf{e}_i$'s are normalized, $\mathbf{e}_i^\top \mathbf{e}_i = \|\mathbf{e}_i\| = 1$.

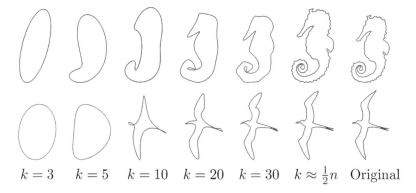

$k = 3 \quad k = 5 \quad k = 10 \quad k = 20 \quad k = 30 \quad k \approx \frac{1}{2}n \quad$ Original

Figure 5.7: Shape reconstruction via Laplacian-based spectral analysis.

In Figure 5.7, we show some results of this type of shape reconstruction with varying k for the seahorse and the bird shapes. As more and more high-frequency spectral coefficients are removed, i.e., with decreasing k, we obtain smoother and smoother reconstructed contours. How effectively a 2D shape can be compressed this way may be visualized by plotting the spectral transform coefficients, the $\tilde{x}_i$'s in Equation (5.6), as done in Figure 5.8. In the plot, the horizontal axis represents eigenvalue indices, $i = 1, \ldots, n$, which roughly correspond to frequencies. One can view the magnitude of the $\tilde{x}_i$'s as the energy of the input signal $\mathbf{x}$ at different frequencies.

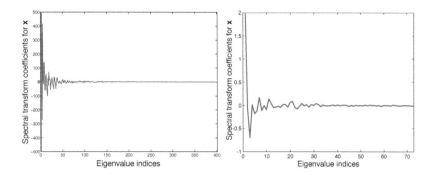

Figure 5.8: Plot of spectral transform coefficients for the x component of a contour; refer to Equation (5.6): seahorse (left) and bird (right). The models are shown in Figure 5.7.

A signal whose energies are concentrated in the low-frequency end can be effectively compressed at a high compression rate, since as a consequence, the total energy at the high-frequency end, representing the reconstruction error, is very low. Such signals exhibit fast decay in their spectral coefficients. Both the seahorse and the bird models contain noisy or sharp features so they are not as highly compressible as a shape with smoother boundaries. This can be observed from the plots in Figure 5.8. Nevertheless, at 2:1 compression ratio, we can obtain a fairly good approximation, as one can see in Figure 5.7.

Filtering and Laplacian smoothing. Compression by truncating the vector $\tilde{\mathbf{x}}$ of spectral transform coefficients can be seen as a *filtering* process. When a discrete filter function f is applied to $\tilde{\mathbf{x}}$, we obtain a new coefficient vector $\tilde{\mathbf{x}}'$, where $\tilde{x}'_i = f(i)\,\tilde{x}_i$, for all i. The filtered signal $\mathbf{x}'$ is then reconstructed from $\tilde{\mathbf{x}}'$ by $\mathbf{x}' = E\,\tilde{\mathbf{x}}'$, where E is the matrix of eigenvectors as defined in Equation (5.5). We next show that Laplacian smoothing is one particular filtering process. Specifically, when we apply the Laplacian smoothing operator S to a coordinate vector m times, the resulting coordinate vector becomes

$$
\begin{aligned}
\mathbf{x}^{\langle m \rangle} = S^m \mathbf{x} = \left(I - \frac{1}{2}L\right)^m \mathbf{x} &= \sum_{i=1}^{n} \left(I - \frac{1}{2}L\right)^m \mathbf{e}_i\,\tilde{x}_i \\
&= \sum_{i=1}^{n} \mathbf{e}_i \left(1 - \frac{1}{2}\lambda_i\right)^m \tilde{x}_i\,.
\end{aligned}
\tag{5.8}
$$

The last equality above results from the definition of eigenvalue/
eigenvector, $L\mathbf{e}_i = \lambda_i \mathbf{e}_i$. For example, we note that by applying
the definition k times, we get $L^k \mathbf{e}_i = \lambda_i^k \mathbf{e}_i$.

Equation (5.8) provides a characterization of Laplacian smooth-
ing in the spectral domain via filtering by $f(\lambda) = (1 - \frac{1}{2}\lambda)^m$. A
few such filters with different m are plotted in the first row of
Figure 5.9. The corresponding Laplacian smoothing leads to at-
tenuation of the high-frequency content of the signal and hence
achieves denoising or smoothing, as shown in the second row.

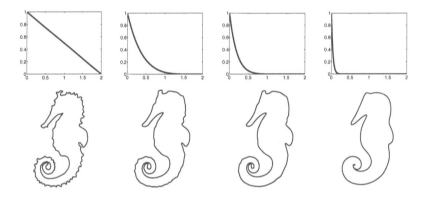

Figure 5.9: First row: filter plots, $\left(1 - \frac{1}{2}\lambda\right)^m$ with $m = 1, 5, 10, 50$. Second row: corresponding results of Laplacian smoothing on the seahorse.

To examine the limit behavior of Laplacian smoothing, we look
at Equation (5.8). It can be shown (via Gerschgorin's theorem
[TB97]) that the eigenvalues of the Laplacian operator are in
the interval $[0, 2]$ and the smallest eigenvalue is $\lambda_1 = 0$. Since
$\lambda \in [0, 2]$, the filter function $f(\lambda) = \left(1 - \frac{1}{2}\lambda\right)^m$ is bounded by the
unit interval $[0, 1]$ and attains the maximum $f(0) = 1$ at $\lambda = 0$. As
$m \to \infty$, all the terms in the right-hand side of Equation (5.8) will
vanish except for the first, which is given by $\mathbf{e}_1 \tilde{x}_1$. Since $\mathbf{e}_1$, the
eigenvector corresponding to the zero eigenvalue, is a normalized,
constant vector, we have $\mathbf{e}_1 = \left[\frac{1}{\sqrt{n}}, \frac{1}{\sqrt{n}}, \cdots, \frac{1}{\sqrt{n}}\right]^\top$. Now taking the
y component into consideration, we get the limit point for Lapla-
cian smoothing as $\frac{1}{\sqrt{n}}[\tilde{x}_1, \tilde{y}_1]$. Finally, noting that $\tilde{x}_1 = \mathbf{e}_1^\top \mathbf{x}$ and
$\tilde{y}_1 = \mathbf{e}_1^\top \mathbf{y}$, we conclude that the limit point of Laplacian smoothing
is the *centroid* of the set of original contour vertices.

Spectral transform vs. Fourier transform. So far we have
started with a purely geometric treatment of contour smoothing

based on 1D discrete Laplacians and then looked at the problem from a signal processing perspective and applied the algebraic tool of spectral analysis to interpret compression, Laplacian smoothing, and signal filtering. For those familiar with signal processing and Fourier analysis, a connection appears to be obvious. In particular, by looking at the plots of the eigenvectors of the 1D discrete Laplacian operator (Equation (5.3)) in Figure 5.6, one notices their resemblance to the sinusoidal curves of the Fourier or DCT basis functions. We now make that connection explicit.

Typically, one introduces discrete Fourier transform (DFT) into the context of Fourier series expansion. Given a discrete signal $\mathbf{x} = [x_1, x_2, \cdots, x_n]^{\top}$, its DFT is given by

$$\tilde{\mathbf{x}}(k) = \frac{1}{n} \sum_{j=1}^{n} \mathbf{x}(k) \, e^{-i\,2\pi(k-1)(j-1)/n} \,, \quad k = 1, \ldots, n \,.$$

The corresponding inverse DFT is given by

$$\mathbf{x}(j) = \sum_{k=1}^{n} \tilde{\mathbf{x}}(k) \, e^{i\,2\pi(j-1)(k-1)/n} \,, \quad j = 1, \ldots, n \,,$$

or

$$\mathbf{x} = \sum_{k=1}^{n} \tilde{\mathbf{x}}(k) \, \mathbf{g}_k \,, \quad \text{where } \mathbf{g}_k(j) = e^{i\,2\pi(j-1)(k-1)/n} \,, \ k = 1, \ldots, n \,.$$

We see that in the context of DFT, the signal $\mathbf{x}$ is expressed as a linear combination of the complex exponential DFT basis functions, the $\mathbf{g}_k$'s. The coefficients are given by the $\tilde{\mathbf{x}}(k)$'s, which form the DFT of $\mathbf{x}$. Fourier analysis, provided by the DFT in the discrete setting, is one of the most important topics in mathematics and has wide-ranging applications in many scientific and engineering disciplines. For a systematic study of the subject, we refer the reader to the classic text by Bracewell [Bra99].

The connection we seek, between DFT and spectral analysis with respect to the Laplacian, is that the DFT basis functions, the $\mathbf{g}_k$'s, form a set of eigenvectors of the 1D discrete Laplacian operator L, given in Equation (5.3). A proof of this fact can be found in Jain's classic text on image processing [Jai89], where a stronger claim with respect to circulant matrices is made. A matrix is circulant if each row can be obtained as a shift (with circular wrap-around) of the previous row. It is clear that L is circulant.

Specifically, let us sort the eigenvalues of L in ascending order,

$$\lambda_k = 2\sin^2 \frac{\pi \lfloor k/2 \rfloor}{n}, \quad k = 2, \dots, n. \tag{5.9}$$

The first eigenvalue λ_1 is always 0. Every eigenvalue of L, except for the first, and possibly the last, has a multiplicity of 2. That is, it corresponds to an eigensubspace spanned by two eigenvectors. If we define the matrix G of the DFT basis as $G_{k,j} = e^{i\,2\pi(j-1)(k-1)/n}$, $1 \le k, j \le n$, then the first column of G is an eigenvector corresponding to λ_1 and the k-th and $(n+2-k)$-th columns of G are two eigenvectors corresponding to λ_k, for $k = 2, \dots, n$. The set of eigenvectors of L is not unique. In particular, it has a set of real eigenvectors as shown in Figure 5.6.

We remark that increasing eigenvalues of L, given in Equation (5.9), correspond to increasing frequencies of the associated eigenvectors. Generally, the frequency of a 1D signal can be measured by the number of zero-crossings. In the case of the DFT basis, the k-th and $(n+2-k)$-th bases both have $2(k-1)$ zero-crossings. For a different set of eigenvectors of L (corresponding to the same set of eigenvalues), such as the ones shown in Figure 5.6, the number of zero-crossings changes but the relationship between eigenvalues and frequencies is still monotonic.

It turns out that the above connection also applies to the DCT and the discrete sine transform, and the three transforms can be treated in a unified manner [Jai89]. Consider the following parametric family of matrices,

$$J = J(k_1, k_2, k_3) = \begin{bmatrix} 1 - k_1\alpha & -\alpha & 0 & 0 & \dots & 0 & k_3\alpha \\ -\alpha & 1 & -\alpha & 0 & \dots & 0 & 0 \\ 0 & -\alpha & 1 & -\alpha & 0 & \dots & 0 \\ \dots & \dots & \dots & \dots & \dots & \dots & \dots \\ 0 & \dots & 0 & -\alpha & 1 & -\alpha & 0 \\ 0 & \dots & 0 & 0 & -\alpha & 1 & -\alpha \\ k_3\alpha & 0 & \dots & 0 & 0 & -\alpha & 1 - k_2\alpha \end{bmatrix}.$$

It can be shown that the basis vectors of the DCT, the discrete sine transform and DFT are, respectively, the complete and orthogonal sets of eigenvectors of $J(1,1,0)$, $J(0,0,0)$ and $J(1,1,-1)$. These eigenvectors all exhibit sinusoidal behavior. The DCT generally possesses better energy compaction properties compared to the other two transforms. This means that for the same signal, the DCT coefficients exhibit faster decay, leading to better compression. The DCT, DFT and discrete sine transforms all admit fast constructions, in time $\mathcal{O}(n \log_2 n)$, where n is the length of the signal, making them practical to use for a variety of signal and image processing tasks.

Based on the above observation, one way to extend Fourier analysis to the surface setting, where our signal will represent the geometry of the surfaces, is to define appropriate discrete Laplacian operators for surface meshes and rely on eigenvectors of the Laplacians to perform Fourier-like analysis. This observation was made by Taubin in his seminal paper on a signal processing approach to mesh smoothing in 1995 [Tau95]. This research has inspired a series of works on spectral mesh processing, as covered in the comprehensive survey by Zhang et al. [ZvKD10].

Spectral analysis on meshes

We now extend spectral analysis of 1D signals presented earlier to surfaces modeled by triangle meshes. A triangle mesh is a tessellation of a surface in 3D using triangles; it is formed by a set of triangles pasted along their edges. The main issue to resolve is the irregularity of the neighborhood of a mesh vertex. In our 1D case, the samples are uniformly spaced, and each vertex has precisely two neighbors. For a mesh, each vertex can have any number of neighbors. In addition, we can no longer ensure a totally uniform spacing between the samples (vertices) over which the mesh signal is defined. Extending the signal representation to the mesh case requires us to define an appropriate *mesh Laplacian* operator. Once that is done, spectral analysis of a mesh signal with respect to the eigenvectors of the mesh Laplacian works in exactly the same way as for its 1D counterpart.

Signal representation and mesh Laplacian. A triangle mesh with n vertices is represented as $\mathcal{M} = (\mathcal{G}, P)$, where $\mathcal{G} = (\mathcal{V}, \mathcal{E})$ models the mesh graph, with $\mathcal{V}$ denoting the set of mesh vertices and $\mathcal{E} \subseteq \mathcal{V} \times \mathcal{V}$ the set of edges, and $P \in \mathbb{R}^{n \times 3}$ represents the geometry of the mesh, given by a matrix of 3D vertex coordinates. Each vertex $i \in \mathcal{V}$ has an associated position vector, denoted by

$\mathbf{p}_i = [x_i, y_i, z_i]$; it corresponds to the i-th row of P. Any function defined on the mesh vertices can be seen as a discrete mesh signal. Here we focus on P, the coordinate signal of the mesh. As before, we will only deal with the x component, denoted by $\mathbf{x}$. The treatment of the y and z components is similar.

Discrete mesh Laplacian operators are linear operators that act on discrete signals defined on the vertices of a mesh. If a mesh $\mathcal{M}$ has n vertices, then the mesh Laplacian will be given by an $n \times n$ matrix. Loosely speaking, a mesh Laplacian operator locally takes a weighted average of the differences between the value of a signal at a vertex and its value at the first-order or immediate neighbor vertices. Although there are several ways to define a mesh Laplacian [ZvKD10], we focus our attention on the simplest of them, which is a direct generalization of the 1D discrete Laplacian from earlier. Specifically, the midpoint between two points adjacent to the point in question in the 1D case is replaced by the centroid of the point's immediate neighbors in the mesh.

Let us denote by W the adjacency matrix of the mesh graph $\mathcal{G}$. Thus

$$W_{i,j} = \begin{cases} 1 & \text{if } (i,j) \in \mathcal{E}, \\ 0 & \text{otherwise}. \end{cases}$$

The degree matrix D is a diagonal matrix of the row sums of W,

$$D_{i,j} = \begin{cases} d_i = |\mathcal{N}(i)| & \text{if } i = j, \\ 0 & \text{otherwise}, \end{cases}$$

where $\mathcal{N}(i)$ is the set of immediate neighbor vertices of i, and d_i is said to be the degree of vertex i. Note that the above W and D matrices are $n \times n$ matrices, where $n = |\mathcal{V}|$.

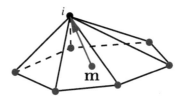

Figure 5.10: Discrete mesh Laplacian (blue arrow) at vertex i. The point $\mathbf{m}$ is the centroid of the immediate neighbors of vertex i.

We define our *mesh Laplacian* matrix T as

$$T = I - D^{-1}W. \tag{5.10}$$

Applying T to the mesh geometry yields the discrete Laplacian at each vertex:

$$\delta(x_i) = x_i - \frac{1}{|\mathcal{N}(i)|} \sum_{j \in \mathcal{N}(i)} x_j, \qquad (5.11)$$

and the same for the y and z components of the vertex. We see that the Laplacian is a vector directed from the centroid of the immediate neighbors of vertex i to vertex i; see Figure 5.10.

Spectral mesh transform. With the mesh Laplacian operator T defined, the way to define the spectral transform of a mesh signal $\mathbf{x}$ with respect to T is exactly the same as the 1D case for L that we saw earlier. Denote by $\mathbf{e}_1, \mathbf{e}_2, \ldots, \mathbf{e}_n$ the normalized eigenvectors of T, corresponding to eigenvalues $\lambda_1 \leq \lambda_2 \leq \cdots \leq \lambda_n$, and let E be the matrix whose columns are the eigenvectors. The vector of spectral transform coefficients $\tilde{\mathbf{x}}$ is obtained by $\tilde{\mathbf{x}} = E^\top \mathbf{x}$. And for each i, we obtain the spectral coefficient by $\tilde{x}_i = \mathbf{e}_i^\top \mathbf{x}$, i.e., via projection.

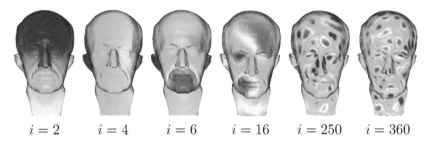

$i = 2 \qquad i = 4 \qquad i = 6 \qquad i = 16 \qquad i = 250 \qquad i = 360$

Figure 5.11: Color plots of a few eigenvectors (corresponding to the i-th smallest eigenvalues) of the mesh Laplacian of the Max Planck model.

In Figure 5.11, we plot several eigenvectors of a mesh model using color coding, where eigenvector entries are used to index into a color map. The oscillatory and low-to-high frequency vibration patterns of the eigenvectors as the eigenvalues increase are evident. In Figure 5.12, we show spectral reconstruction, as defined in Equation (5.7), of a mesh model with progressively more spectral coefficients added. As we can see, higher-frequency contents of the geometric mesh signal manifest themselves as rough geometric features over the shape's surface; these features can be smoothed out by taking only the low-frequency spectral coefficients in a reconstruction. A tolerance on such loss of geometric features leads to a JPEG-like compression of mesh geometry [KG00].

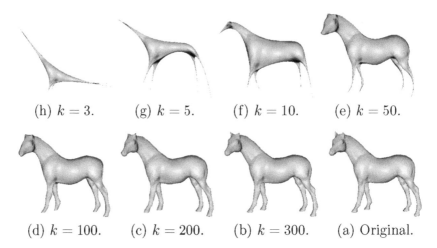

(h) $k = 3$.   (g) $k = 5$.   (f) $k = 10$.   (e) $k = 50$.

(d) $k = 100$.   (c) $k = 200$.   (b) $k = 300$.   (a) Original.

Figure 5.12: Shape reconstruction via spectral analysis using a mesh Laplacian operator, where k is the number of eigenvectors or spectral coefficients used. The original model has 7,502 vertices and 15,000 faces.

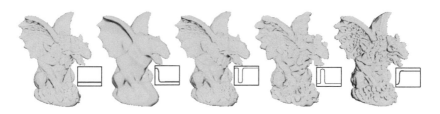

Figure 5.13: A few visual examples of mesh filtering with plots of the applied filters shown on the side.

Just as in the 1D classical case, the oscillatory behavior of the Laplacian eigenvectors and the implied spectral transform of geometric mesh signals allow us to apply Fourier-like analysis to mesh models. In addition to spectral geometry compression (Figure 5.12), any type of filtering can be applied to the spectral coefficients [Tau95, KR05, VL08] to obtain a global editing of a mesh shape. A few examples are shown in Figure 5.13, mimicking their counterparts in the 1D classical case shown in Figure 5.9. A variety of other applications that utilize the spectral coefficients or the eigenvalues and eigenvectors themselves, as derived from different variants of the mesh Laplacian operator (Equation (5.10)), have been developed over the years. We refer the reader to the recent survey [ZvKD10] for a comprehensive coverage.

Comparison to classical Fourier analysis. Comparing spectral mesh transforms to classical Fourier analysis, there are at least two crucial differences. First, while the DFT and DCT basis functions are fixed as long as the length of the signals in consideration is determined, the eigenvectors of the mesh Laplacian T, on the other hand, would change with the mesh graph connectivity. Since the same surface can be tessellated in different ways by a triangle mesh, it may be desirable to use a mesh Laplacian operator that is geometry-aware and not dictated by the particular surface tessellation chosen.

A popular geometric mesh Laplacian is the cotangent operator, introduced to this research domain by Pinkall and Polthier [PP93]. For a closed triangle mesh, it is given by $L^{\mathrm{cot}} = D^{\mathrm{cot}} - W^{\mathrm{cot}}$, where W^{cot} is defined by the cotangent weights:

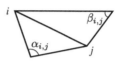

$$W_{i,j}^{\mathrm{cot}} = \frac{1}{2}\left(\cot \alpha_{i,j} + \cot \beta_{i,j}\right),$$

where $\alpha_{i,j}$ and $\beta_{i,j}$ are opposite angles to edge (i,j) in the mesh and if $(i,j) \notin \mathcal{E}$ (i.e., there is no such edge in the mesh), $W_{i,j}^{\mathrm{cot}} = 0$. The matrix D^{cot} is a diagonal matrix of the row sums of W^{cot}. Since the cotangent weights are meant to model mesh geometry, the mesh Laplacian L^{cot} and the corresponding spectral transform are shape-dependent and thus insensitive to change of mesh tessellation, unlike T.

The second difference between classical DFT and spectral mesh transforms is in the computational aspect. While the classical DFT admits fast $\mathcal{O}(n \log n)$ computations, e.g., via the fast DFT or FFT [Bra99], computing the eigenvectors of general mesh Laplacians, such as T or L^{cot}, has cubic time complexity in the worst case. There are several ways to speed up this process, e.g., via spectral shift and invert [VL08], algebraic multi-grid methods, or settling for approximated results [ZvKD10], but these are beyond the scope of our coverage in this book.

Chapter 6

Solution of Linear Systems

Chen Greif

Very often, the primary computational bottleneck in the solution of a problem in applied geometry and other areas of applications is the numerical solution of one or several linear systems that arise throughout the computation:

$$Ax = b,$$

where $A \in \mathbb{R}^{n \times n}$ and $\mathbf{x}, \mathbf{b} \in \mathbb{R}^n$. Even a modest understanding of the properties of the underlying matrix may help in making a good choice of a solution method.

If A is non-singular, the above linear system has a unique solution for any given right-hand side vector $\mathbf{b}$. Numerical methods for solving linear systems of equations can generally be divided into two classes: *direct methods*, which seek the exact solution within a finite number of steps, and *iterative methods*, which are based on starting with an initial guess and iterating until the approximate solution is sufficiently close to the exact solution by an acceptable measure. The quality of a numerical method is determined by the speed of computation, the computer storage requirements, the expected accuracy of the numerical solution, and other considerations. The dimensions of the problem, sparsity of the underlying matrix, symmetry, positive definiteness, and so on, are critical factors.

Gaussian elimination and LU decomposition

Direct solution methods are based on the principle that we can simplify a linear system without changing its solution, by elementary row operations. To that end, we aim to zero out all elements below the main diagonal, one by one, by performing row operations until we obtain an upper triangular system. Once such a system is obtained, it is trivial to solve it by backward substitution.

This *Gaussian elimination* procedure implicitly decomposes A into a product of a unit lower triangular matrix L and an upper triangular matrix U. Together with the backward substitution, the entire solution algorithm for $A\mathbf{x} = \mathbf{b}$ can therefore be described in three steps, the first two of which are usually carried out simultaneously and implicitly:

1. *LU decomposition*: $A = LU$;

2. *Forward substitution*: solve $L\mathbf{y} = \mathbf{b}$;

3. *Backward substitution*: solve $U\mathbf{x} = \mathbf{y}$.

In practice, given a general dense matrix A, it is not necessary to allocate additional memory for the two matrices L and U; we could instead overwrite A. This is illustrated in Algorithm 1. Similarly, $\mathbf{b}$ may be overwritten by $L^{-1}\mathbf{b}$ during the first two steps of the three-step procedure.

Forming the LU decomposition takes $\mathcal{O}(n^3)$ floating point operations for a general dense $A \in \mathbb{R}^{n \times n}$. It takes $\mathcal{O}(n^2)$ floating point operations to perform backward or forward substitutions.

Algorithm 1 The LU decomposition of a matrix A. Upon exit, the entries of A have been overwritten with the entries of L (below the main diagonal) and the entries of U (main diagonal and above). The diagonal entries of L are all equal to 1.

> **for** $k = 1, \ldots, n - 1$ **do**
>> **for** $i = k + 1, \ldots, n$ **do**
>>> $a_{i,k} = \frac{a_{i,k}}{a_{k,k}}$
>>> **for** $j = k + 1, \ldots, n$ **do**
>>>> $a_{i,j} = a_{i,j} - a_{i,k}a_{k,j}$
>>> **end for**
>> **end for**
> **end for**

Pivoting. The process of Gaussian elimination may easily break down. For example, if $a_{1,1} = 0$, then in Algorithm 1 there will be a division by zero immediately upon attempting to replace $a_{2,1}$ by $\frac{a_{2,1}}{a_{1,1}}$ (see third line of the algorithm). Even if $a_{1,1}$ is not equal to zero, but is a small number, say $a_{1,1} = \varepsilon \ll 1$, we may still get in trouble, since dividing by ε may result in a very large number that could cause harmful inaccuracy. Gaussian elimination will thus fail or perform poorly.

A way to fix this is called *partial pivoting*. Before we try to zero out elements below a given diagonal element (or *pivot*), we first swap rows and put the largest element (in absolute value) in a given column on the diagonal. This will prevent division by zero or by a very small number when the matrix is non-singular and reasonably well conditioned. The procedure can thus be described as an LU decomposition that allows for permutations, and can be written as $PA = LU$, where P is a permutation matrix.

One can find examples where Gaussian elimination with partial pivoting (GEPP) fails, but they are rare. It is a robust procedure in general, and is the most commonly used technique for solving general dense linear systems.

In MATLAB, given a matrix A and a vector $\mathbf{b}$, one can solve the associated linear system $A\mathbf{x} = \mathbf{b}$ by `x=A\b`. The backslash operation is very effective, and by default it solves the linear system using GEPP. If we want to obtain the LU decomposition, then the command `[L,U,P]=lu(A)` will give us the L and U factors, and the permutation matrix P that indicates the row swaps that have been carried out in the process of pivoting.

Cholesky decomposition. The choice of an effective solution method depends also on the numerical properties of the matrix. One prominent property that leads to a special solution method is *symmetric positive definiteness.* Recall that a matrix A is symmetric positive definite (SPD) if $A = A^{\top}$ and $\mathbf{x}^{\top} A \mathbf{x} > 0$ for any nonzero vector $\mathbf{x} \neq \mathbf{0}$. SPD matrices allow for fast and robust solution methods, and pivoting is not required (at least theoretically). For an SPD matrix, the LU decomposition reduces to the well-known *Cholesky decomposition*:

$$A = FF^{\top},$$

where F is lower triangular. The fact that the upper triangular factor is merely the transpose of the lower triangular factor follows from symmetry and yields significant savings. The Cholesky decomposition algorithm is typically numerically stable. It is given in Algorithm 2.

Algorithm 2 The Cholesky decomposition of a symmetric positive definite matrix A. Upon exit, the entries of A on its diagonal and below it have been overwritten with the entries of the lower triangular Cholesky factor F.

---

for $k = 1, \ldots, n$ **do**
  $a_{k,k} = \sqrt{a_{k,k}}$
  for $i = k + 1, \ldots, n$ **do**
    $a_{i,k} = \frac{a_{i,k}}{a_{k,k}}$
  end for
  for $j = k + 1, \ldots, n$ **do**
    for $i = j, \ldots, n$ **do**
      $a_{i,j} = a_{i,j} - a_{i,k} a_{k,j}$
    end for
  end for
end for

Exploiting the nonzero structure

In our discussion so far, we have considered the general case where the matrix does not have any particular nonzero structure. For dense linear systems, Gaussian elimination/LU decomposition is the method of choice. Things become more interesting when there is a special structure that we wish to exploit.

The *nonzero structure*, or *sparsity pattern*, is an important consideration. In general, there is no need to store zero values; we just record the location and values of the nonzero entries. Having a good idea of the sparsity pattern of the LU factors can often tell us something about how much computational work will be entailed in solving the associated linear system.

Narrow-banded matrices. Narrow-banded matrices are a good and useful example of matrices with a special sparsity pattern. For example, tridiagonal matrices arise frequently in applications.

These are matrices that have precisely three nonzero diagonals: the main diagonal, along with the superdiagonal and subdiagonal immediately above and below the main diagonal, respectively. For tridiagonal matrices, the Gaussian elimination process is significantly cheaper than in the general dense case. The backward and forward substitutions involve only $\mathcal{O}(n)$ operations, and so does the decomposition. So, the entire process of solution takes $\mathcal{O}(n)$ floating point operations rather than $\mathcal{O}(n^3)$ operations. This is a very dramatic difference. Also, the storage required is only $\mathcal{O}(n)$, compared with $\mathcal{O}(n^2)$ in the general case. A similar situation occurs for other matrices that involve a small number of superdiagonals and subdiagonals immediately next to the main diagonal.

Gaussian elimination and sparsity. Things get more delicate when the nonzeros of the matrix do not necessarily appear near the main diagonal, but the overall degree of sparsity is still very high. In particular, if the matrix is narrow-banded but there is some sparsity within the band, or if the matrix is sparse but the location of its nonzeros does not have a clear, structured pattern, this is when we may see a difference in performance among various solution strategies.

A common example is the discrete two-dimensional Laplacian arising from discretization of the Poisson equation on a uniform mesh on a square domain, subject to Dirichlet boundary conditions. The matrix is symmetric positive definite in this case, and its sparsity pattern is given in Figure 6.1. It is a narrow-banded matrix, but within the band we see much sparsity.

Applying the Cholesky decomposition results in some loss of sparsity. The matrix has approximately $5n$ nonzero entries, but the Cholesky factor contains about $n\sqrt{n}$ nonzeros, and the cost of the decomposition is $\mathcal{O}(n^2)$ floating-point operations.

An important family of methods aim at reordering the unknowns in a way that reduces storage requirements and computational work. Popular approaches include reducing the bandwidth or the envelope of the matrix and reducing the expected fill-in in the decomposition. Discussing ordering strategies at length is beyond the scope of this chapter. Notable methods here are reverse Cuthill–McKee, approximate minimum degree, and nested dissection.

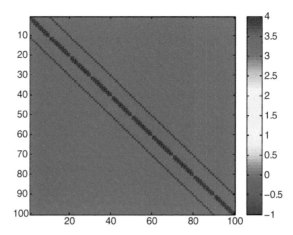

Figure 6.1: Sparsity pattern of a 100×100 discrete two-dimensional Laplacian on a uniform mesh.

We end this part of the discussion with a note of caution. Inverses of matrices are almost always entirely dense. For example, for the Laplacian matrix, we have that the Cholesky factors are narrow-banded, but the inverse is dense; see Figure 6.2. For this reason, there is a fundamental difference between solving a linear system using factorizations, and solving it by computing the inverse of the matrix explicitly.

Iterative methods

Direct methods are extremely useful, and we may even dare say that if you can apply them conveniently (that is, without waiting too long for the computer to complete the computation and without running out of memory) you should use them. However, their drawbacks may in some situations prompt us to consider a different approach. At the top of the list of things to worry about is the fact that the Gaussian elimination (or LU decomposition) process may cause *fill-in*, i.e., L and U may have nonzero elements in locations where the original matrix A has zeros. The two-dimensional Laplacian is a good example; the Cholesky factor is dense within the band. If the amount of fill-in is significant, then applying the direct method may become costly both in terms of computational work and in terms of storage requirements. A

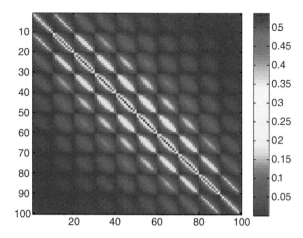

Figure 6.2: An image of the inverse of a 100×100 discrete two-dimensional Laplacian on a uniform mesh. The inverse has entries that strongly decay far from the main diagonal, but it is a fully dense matrix.

good ordering strategy may reduce the fill-in, but its effectiveness is often limited.

Another issue with direct methods is that they cannot make good use of an initial guess. Imagine, for example, that you are solving a geometric problem related to tracking the evolution of a certain image in time, and there is much in common between two consecutive images. You might then wish to use the image in the previous time step as a convenient starting point for computation of the solution at the current time step. Direct methods are not easily conducive to such a philosophy.

Finally, sometimes the matrix of a linear system is not available explicitly. It may be available only as some kind of a black box: an operator that gives you $A\mathbf{x}$ if you provide it with $\mathbf{x}$. This can happen in several areas of applications, either due to the nature of the problem or due to a huge size of the linear system, which does not allow for storing the matrix conveniently.

Iterative methods are based on the idea that, using an initial guess and essentially a sequence of matrix–vector products, we invert an easy matrix several times rather than inverting A once. We seek an approximation to A (in a sense that will become clear below) and keep iterating, computing an approximate solution repeatedly until we are satisfied.

Our general assumption henceforth will be that we have a large and sparse linear system. These days, *large* means at least millions of unknowns, if not more.

Basic stationary methods

Given $A\mathbf{x} = \mathbf{b}$, suppose $A = M - N$. We have $M\mathbf{x} = N\mathbf{x} + \mathbf{b}$, which may lead to the iteration

$$M\mathbf{x}_{k+1} = N\mathbf{x}_k + \mathbf{b}.$$

This is a *stationary* or a *fixed-point* iteration.

A few questions arise at this point. One is, what should M be? Another is, why and when do we expect this process to converge to the solution. And a third question is, even if there is convergence, how fast is it?

Let us address the first question: the choice of M. It should satisfy two contradictory requirements: on the one hand, it should be significantly easier to invert than A; on the other hand, M should be close to A in a certain way, otherwise it will take too many iterations to converge.

The most basic stationary methods tend to better satisfy the first aspect of the above two: ease of inversion. Suppose A is additively split into its diagonal D, its strictly upper triangular part E and its strictly lower triangular part F, so that $A = D + E + F$. Then the following schemes are the most basic and well-known stationary methods:

- Jacobi: $M = D$;

- Gauss–Seidel: $M = D + E$.

To illustrate how the iterations are done, consider the Jacobi scheme. We choose $M = D$, which gives $N = -(E + F)$. Given an initial guess $\mathbf{x}_0$, the k-th iteration is

$$\mathbf{x}_{k+1} = -D^{-1}[(E + F)\mathbf{x}_k + \mathbf{b}].$$

Jacobi and Gauss–Seidel are *terribly slow* methods. There are nice ways to improve on them, for example a method called Successive Over Relaxation (SOR), which is an improved Gauss–Seidel.

This method is significantly faster than Gauss–Seidel, but it requires knowing the optimal value of a certain parameter, and this inserts a difficulty into the process.

Altogether, stationary schemes do not excel in terms of fast convergence, but they are useful because they are extremely simple to program and for easy enough problems, they can be sufficiently effective. In addition to the simplicity of these methods, convergence analysis is easy to perform and can give an idea of how hard it is to solve the linear system.

The Jacobi method is very slow but it is highly parallelizable, and there are block versions of it that are faster and work well in a parallel environment. Indeed, for Jacobi, each component in the $(k + 1)$-st iterate depends only on components of the k-th iterate. Therefore, all updates can be done simultaneously. This is why, to a certain degree, Jacobi and its variants have been resurrected ever since high-performance computing has become an important paradigm in scientific computing. Gauss–Seidel is not easily parallelizable but it is also useful these days, especially as a *smoother* for multigrid methods.

It can be shown that (asymptotic) convergence is governed by the eigenvalues of the *iteration matrix* $T = M^{-1}N = I - M^{-1}A$. A necessary and sufficient condition for convergence is that the eigenvalues of T are all smaller than 1 in magnitude (note that there could be complex eigenvalues). The smaller the maximal magnitude of the eigenvalues, the faster the convergence. This is true not only for Jacobi or Gauss–Seidel, but for any stationary scheme associated with a splitting $A = M - N$. If some of the eigenvalues of T are very close to 1, then we may experience trouble. Unfortunately, this is often the case in many applications, particularly in the solution of discretized partial differential equations.

Krylov subspace methods

A weakness of stationary schemes is that they do not make use of information that has accumulated throughout the iteration. Modern methods are designed differently, and aim to exploit current information to determine the next step. The conjugate gradient (CG) method is an example of a technique that is based on such a philosophy; it works for symmetric positive definite matrices.

Consider the optimization problem of finding a vector $\mathbf{x}$ that minimizes

$$\phi(\mathbf{x}) = \tfrac{1}{2}\mathbf{x}^\top A\mathbf{x} - \mathbf{b}^\top\mathbf{x}\,,$$

where A is symmetric positive definite. Then it is possible to show that the solution satisfies $A\mathbf{x} = \mathbf{b}$; just differentiate the objective function with respect to $\mathbf{x}$ and you will see this. Therefore, we can adopt optimization techniques. We iterate

$$\mathbf{x}_{k+1} = \mathbf{x}_k + \alpha_k\mathbf{p}_k\,,$$

where the vector $\mathbf{p}_k$ is the *search direction* and the scalar α_k is the *step size*. CG proceeds by defining special search directions and minimizing $\|\mathbf{e}_k\|_A = \sqrt{\mathbf{e}_k^T A\mathbf{e}_k}$, where $\mathbf{e}_k = \mathbf{x} - \mathbf{x}_k$ is the error. (Note that if A is not SPD, this norm, and hence the minimization problem, is not well defined.) The search directions are A-conjugate, which means that they are orthogonal with respect to an A-weighted inner product. CG is given in Algorithm 3.

Algorithm 3 The conjugate gradient algorithm

Given $\mathbf{x}_0$, compute $\mathbf{r}_0 = \mathbf{b} - A\mathbf{x}_0$ and set $\mathbf{p}_0 = \mathbf{r}_0$
for $k = 1$ until convergence **do**
   $\mathbf{s}_{k-1} = A\mathbf{p}_{k-1}$
   $\alpha_{k-1} = \dfrac{\mathbf{r}_{k-1}^\top\mathbf{r}_{k-1}}{\mathbf{p}_{k-1}^\top\mathbf{s}_{k-1}}$
   $\mathbf{x}_k = \mathbf{x}_{k-1} + \alpha_{k-1}\mathbf{p}_{k-1}$
   $\mathbf{r}_k = \mathbf{r}_{k-1} - \alpha_{k-1}\mathbf{s}_{k-1}$
   $\beta_{k-1} = \dfrac{\mathbf{r}_k^\top\mathbf{r}_k}{\mathbf{r}_{k-1}^\top\mathbf{r}_{k-1}}$
   $\mathbf{p}_k = \mathbf{r}_k + \beta_{k-1}\mathbf{p}_{k-1}$
end for

The conjugate gradient method is a member of the well-known family of *Krylov subspace methods*. The idea of these methods is to seek a solution within the following subspace:

$$\mathbf{x}_k \in \mathbf{x}_0 + \mathcal{K}^k(A; \mathbf{r}_0) \equiv \mathbf{x}_0 + \mathrm{span}\{\mathbf{r}_0, A\mathbf{r}_0, A^2\mathbf{r}_0, \ldots, A^{k-1}\mathbf{r}_0\}\,,$$

where $\mathbf{x}_0$ is the initial guess and $\mathbf{r}_0 = \mathbf{b} - A\mathbf{x}_0$ is the initial residual.

This is done as follows:

- We form an orthogonal basis for the Krylov subspace. This is known as the *Arnoldi process*. After k steps, the procedure generates a decomposition of the form

$$AV_k = V_{k+1}H_{k+1,k},$$

where V_k is an n-by-k matrix with orthonormal columns that form the basis for the Krylov subspace, and $H_{k+1,k}$ is an upper Hessenberg matrix (a matrix that looks like an upper triangular matrix plus one additional subdiagonal immediately below the main diagonal) that contains coordinates of the basis vectors with respect to the matrix A. When the matrix is symmetric, the upper Hessenberg reduces to a tridiagonal matrix and the procedure simplifies; this is known as the *Lanczos algorithm*. The Arnoldi process is given in Algorithm 4.

- We then seek an approximate solution within the subspace that satisfies an optimality condition that makes sense. As we said previously, CG minimizes $\|\mathbf{e}_k\|_A$. Another example for a sensible condition would be to require that the norm of the residual, $\|\mathbf{b} - A\mathbf{x}_k\|_2$, is minimal over the subspace. This leads to well-known methods such as MINRES or GMRES.

Algorithm 4 The Arnoldi algorithm

Given $\mathbf{r}_0 = \mathbf{b} - A\mathbf{x}_0$, compute $\mathbf{v}_1 = \mathbf{r}_0/\|\mathbf{r}_0\|_2$
for $j = 1$ to k **do**
  $\mathbf{z} = A\mathbf{v}_j$
  for $i = 1$ to j **do**
    $h_{i,j} = \mathbf{v}_i^\top \mathbf{z}$
    $\mathbf{z} \leftarrow \mathbf{z} - h_{i,j}\mathbf{v}_i$
  end for
  $h_{j+1,j} = \|\mathbf{z}\|_2$
  if $h_{j+1,j} = 0$ **then**
    quit
  end if
  $\mathbf{v}_{j+1} = \mathbf{z}/h_{j+1,j}$
end for

Preconditioning

Convergence rates of nonstationary solvers typically depend on two factors:

- distribution of the eigenvalues of the matrix;

- the condition number of the matrix.

If the matrix is very ill-conditioned, or if its eigenvalues are distributed in a non-favorable fashion, convergence may be slow. Since A is given and is beyond our control, one possible idea for overcoming these difficulties is to define a *preconditioner* M such that one or both of the above two properties are better for $M^{-1}A$, and solve $M^{-1}A\mathbf{x} = M^{-1}\mathbf{b}$ rather than $A\mathbf{x} = \mathbf{b}$. This is called *left preconditioning* because M^{-1} is multiplied on the left of the original linear system. *Right preconditioning* amounts to forming $AM^{-1}\mathbf{y} = \mathbf{b}$ and then solving $\mathbf{x} = M^{-1}\mathbf{y}$. There is also *split preconditioning*, which is based on solving $M_1^{-1}AM_2^{-1}\mathbf{y} = M_1^{-1}\mathbf{b}$ and $\mathbf{x} = \mathbf{M}_2^{-1}\mathbf{y}$. There are some theoretical differences among the different types of preconditioning just described, as each of them gives rise to a slightly different Krylov subspace.

To produce an effective method, the preconditioner M must be easily invertible. At the same time, it is desirable to have at least one of the following properties hold: the condition number of $M^{-1}A$ is significantly smaller than that of A, and/or the eigenvalues of $M^{-1}A$ are much better clustered compared to those of A. The act of finding a matrix M that makes $M^{-1}A$ spectrally superior to A but without being too close to A in terms of difficulty of inversion is delicate and challenging.

There are several widely used preconditioning approaches:

- Stationary preconditioners, such as Jacobi, Gauss–Seidel, SOR;

- incomplete LU factorizations;

- multigrid and multilevel preconditioners;

- preconditioners tailored to the problem in hand, that rely for example on the properties of the underlying differential operators.

Multigrid is based on a two-step recursive procedure that dampens rough modes of the error by applying a smoother, and dampens smooth modes of the error by projecting the problem onto a coarser mesh. It is a very effective method; in fact, it has linear complexity for certain problems such as the discrete Poisson equation. In recent years, multigrid has been used not only as a stand-alone solver but also as a preconditioner. Programming a multigrid algorithm is delicate at times, but the results are often impressive, and multigrid is popular especially for problems involving differential equations.

Incomplete LU factorizations are based on the simple yet elegant idea of approximating the LU factors of A by sparse matrices. They come in two main flavors. *Static-pattern* incomplete LU factorizations are based on determining ahead of time the sparsity patterns of the approximate factors, and working toward making the nonzeros of these factors equal to the corresponding entries of A. This approach is simple, though it is not always effective.

A more sophisticated approach is that of *dynamic-pattern* incomplete LU factorizations. This approach is based on dropping entries whenever they fall below a prescribed threshold, commonly denoted by τ, regardless of their location in the matrix (an exception to this are nonzero diagonal entries, which are always kept). To make sure no excessive memory resources are used, another parameter (commonly denoted by p) determines how many entries in each row are to be kept; only the largest p are stored. The ILUTP method, which is based on this philosophy and also involves pivoting, is very popular. A disadvantage of incomplete factorizations is that they cannot easily exploit a specific structure in the matrix.

Let us illustrate the dramatic effect of preconditioning. In MATLAB, the unpreconditioned way of applying the conjugate gradient method may be implemented as follows:

```
[x,flag,relres,iter,resvec]=pcg(A,b,1e-8,200);
```

Input is the matrix A, the right-hand side **b**, a convergence tolerance (`1e-8` in this case) and a maximal iteration count (200 in this case). As output, we obtain the solution x, a flag that tells us whether convergence was obtained (0 if everything is OK), the relative residual `relres`, the iteration count `iter` and the history of norm of residual, `resvec`.

Suppose we now wish to apply a preconditioner. A natural choice would be the incomplete Cholesky decomposition with zero fill-in (static pattern). Here is an example of how we may incorporate this preconditioning approach into the solver:

```
G=ichol(A);
[x,flag,relres,iter,resvec]=pcg(A,b,1e-8,200,G,G');
```

Figure 6.3 illustrates the convergence of the preconditioned conjugate gradient scheme. After 200 iterations, the un-preconditioned

CG method is not even close to convergence, and the residual is still at a level of approximately 0.01. In contrast, 100 iterations of preconditioned CG are sufficient to reduce the residual from an initial level of approximately 10^2 to 10^{-6}, i.e., by a factor of 10^{-8}.

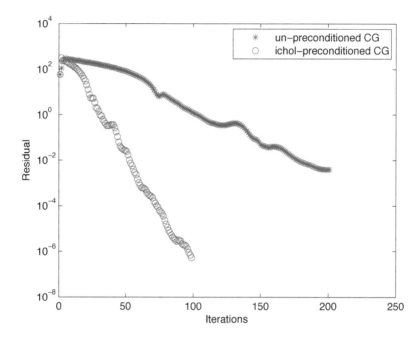

Figure 6.3: Convergence of the preconditioned conjugate gradient method for the $10,000 \times 10,000$ discrete Laplacian. Incomplete Cholesky was used as a preconditioner.

What method should you choose?

With such a wealth of available methods, it is natural to ask how to go about choosing an appropriate method. The answer to such a question is not unique, but let us try to provide some general guidelines that are reasonably within the general consensus.

- **If your matrix is dense and/or small.** Gaussian elimination is generally very robust and reliable, and if the matrix is dense (i.e., its number of nonzeros is close to n^2) and/or small, GEPP is typically the method of choice. There is no real reason to consider using an iterative method in such cases.

- **If your matrix is large and sparse, or not available explicitly.** Matrices that are sparse and very large in size cannot be easily decomposed. Gaussian elimination preceded by sophisticated ordering techniques may do well, but not always. In such situations, an iterative method may work effectively. Also, there are applications where the matrix is only available implicitly: for a given vector $\mathbf{z}$, a routine that provides the matrix–vector product $A\mathbf{z}$ is provided. This is also a case where an iterative method would be a natural choice.

- **If your matrix is narrow-banded.** The critical question here would be whether the matrix is sparse within the band. If it is dense within the band or if the bandwidth is small, for example if the matrix is tridiagonal or pentadiagonal, then Gaussian elimination will work well, with linear complexity, and is your best bet. But if the matrix is sparse within the band—the discrete Laplacian is such an example—and Gaussian elimination with a sophisticated ordering strategy is not sufficiently effective, then it is worth considering iterative methods, and then the choice of the iterative method depends on the type of matrix: see below.

- **If your matrix is symmetric positive definite, dense and/or small.** Symmetric positive definite matrices form a special class of matrices. If a direct method is to be used, then instead of using GEPP, one should use the Cholesky decomposition.

- **If your matrix is symmetric positive definite, large and sparse.** No question here: conjugate gradient!

- **If your matrix is symmetric indefinite, dense and/or small.** A popular variant of Gaussian elimination in this case is the Bunch–Kaufman algorithm, which generates a factorization of the form

$$PAP^\top = LBL^\top,$$

where P is a permutation matrix (pivoting here is symmetric), L is unit lower triangular, and B is block diagonal with 1×1 and 2×2 blocks.

- **If your matrix is symmetric indefinite, large and sparse.** MINRES is a popular iterative method in this case. It is based on short recurrences that minimize the norm of the residual within the Krylov subspace.

- **If your matrix is nonsymmetric, large and sparse.** The choice of a method here depends on storage requirements. If you are not particularly limited in terms of memory resources, then GMRES is a very effective method: it minimizes the residual within the Krylov subspace and is very robust. Restarted versions of GMRES keep the memory requirements at a cap, at the price of compromising on the minimization property. If, on the other hand, memory is scarce, then it is better to resort to methods such as BiCG-STAB or IDR, which are based on short recurrences but do not satisfy an optimality property. Their convergence behavior may be erratic, but they are attractively economical and are very popular.

All iterative methods without exception should be applied with an appropriate preconditioner. The choice of an appropriate preconditioner is not easy, and we have only given a taste of available preconditioning techniques. These days we are reaching a point where solvers and their theoretical properties are relatively well understood, and it is preconditioning techniques that make the difference in performance. For this reason, the study of preconditioning techniques is important and useful.

Conclusions

We have discussed various solution methods for linear systems. A central factor in the choice of solvers is the sparsity of the underlying matrix. For general dense matrices with no particular structure, Gaussian elimination is very effective. Here we have discussed the LU decomposition that is generated in the process, and the importance of pivoting.

Narrow-banded matrices come in a few flavors, and the main distinction that should be made is between ones that are dense within the band and ones that are sparse within the band. For the former, nothing beats Gaussian elimination. For the latter, sensible ordering strategies may reduce computational costs, but the issue of fill-in is a potentially serious drawback that cannot be easily overcome, and here is where iterative methods have a distinct advantage over direct methods.

We have also discussed other considerations, such as symmetry and positive definiteness. These are important factors in the choice of solvers, and one should always aim to identify whether or not the matrix of the linear system possesses these properties. Special solution approaches, such as the Cholesky factorization (direct) or conjugate gradient method (iterative), are the methods of choice for SPD systems.

Iterative methods may be effective in situations where direct methods do not fare so well. We have described a few such possible situations. Iterative methods are different than direct methods. They do not aim to solve the system within a finite number of steps. Instead, the idea behind them is that of repeatedly correcting the computed solution so that it gets closer to the exact solution. The goal is not necessarily to compute a maximally accurate solution, but rather to compute an approximate solution rather quickly. Indeed, in the iterative paradigm we are typically content with a solution that can be computed within a modest number of steps and is, say, accurate to 6 or 8 decimal digits. State-of-the-art iterative methods involve a rather intricate theory, but they are relatively easy to implement.

There are many software packages out there that offer reliable black-box linear solvers. MATLAB provides a very easy access, using single commands. In addition, repositories like `netlib` at `http://www.netlib.org` provide state-of-the-art and highly reliable solvers.

There are several authoritative books that provide illuminating descriptions of direct and iterative linear solvers. Two such references are Davis [Dav06] for direct methods and Saad [Saa03] for iterative methods.

Chapter 7

Laplace and Poisson

Daniel Cohen-Or and Gil Hoffer

Figure 7.1: Pierre-Simon, Marquis de Laplace (left) and Simon-Denis Poisson (right).

One of the enjoyable experiences of scientists today is solving new problems with mathematical tools that were developed hundreds of years ago. These tools are typically generic and were originally developed to solve problems in physics. In this chapter we make use of the well-known equations of Laplace (Pierre-Simon, Marquis de Laplace 1749–1827) and Poisson (Simon-Denis Poisson 1781–1840). (See Figure 7.1.) The two equations, respectively, have an extremely simple form:

$$\Delta f = 0 \qquad \text{for some function } f \qquad (7.1)$$

and

$$\Delta f = \operatorname{div} \mathbf{g} \qquad \text{for some function } f \text{ and vector field } \mathbf{g} . \quad (7.2)$$

These two equations are partial differential equations and they
have a broad use in diverse branches of mathematical physics.
They are widely used in electromagnetism, astronomy and fluid
dynamics, but in this chapter we will interpret them in the context of image and geometry processing. In the following, we will
show some interesting image editing and geometric problems and
how they can be solved by simple means. We will make use of
these equations, but without using the terminology of differential
equations or physics. The relationship of our basic terminology
to these equations will be made clear later, toward the end of the
chapter.

Operating on derivatives

Traditionally, both in computer graphics and image editing, we
think of the main entities as mathematical functions; surfaces of
3D objects are either parameterized analytically or discretely using
vertices interconnected to form a polygonal mesh. Digital images
are also given over a discrete 2D lattice that samples the color
intensities in the scene. It is very natural to operate on these
functions directly, i.e., explicitly change or set vertex coordinates
when deforming an object or the pixel intensities when editing an
image. In this chapter we explore another possibility to manipulate these types of functions, where the *derivatives* are the basic
handles for the operation. In this approach, the differences between neighboring vertices or pixels are mapped, minimized and
copied and pasted, depending upon the application. We will mention here a few applications that operate on the derivatives and
describe the advantages of performing these operations in the *gradient* domain. But before doing that, we begin by establishing
the basic notation and tools that will assist us, and describe their
relationship to the differential equations we mentioned earlier.

Discrete partial derivatives. Let's start by considering a discrete scalar function $f(x, y) \colon N \times N \to [0, 1]$, defined on a discrete
finite lattice, where x and y are integers; the values of f are real
numbers in, say, the segment $[0, 1]$. In most of the examples we
present here, this function represents the image intensities at each

pixel. One way to define the discrete partial derivatives of f is

$$f_x(x, y) = f(x+1, y) - f(x, y) \quad \text{and} \quad f_y(x, y) = f(x, y+1) - f(x, y). \tag{7.3}$$

If we assume that $f(x, y)$ samples a continuously differentiable function defined on the plane, then it is easy to see, using Taylor expansion, that these formulas approximate that function's first derivatives. This approximation becomes more accurate as the lattice spacing tends to zero or when the higher-order (second and above) derivatives tend to zero. Although digital images are defined at a fixed resolution and they are discontinuous in their essence, we use the terminology of differential operators to describe their discrete analogs. The particular definition in Equation (7.3), used to approximate the partial derivatives, is not the only one possible, but since it avoids certain numerical complications, we will use it throughout our discussion. The vector consisting of these partial derivatives is called the *gradient* vector. We use the same term for the discrete analog, $\big[f_x(x, y), f_y(x, y)\big]$, and also the same notation, $\nabla f(x, y)$. Now, we will describe a basic tool that will allow us to define and alter images indirectly, through their *derivatives*, and *not* by setting their pixel values directly. This can be done in several ways that are better suited for different applications. In this chapter we describe one of these methods, namely, the (homogeneous) Poisson equation.

Poisson equation. The (homogeneous) Poisson equation allows us to reconstruct a function given values for its directional derivatives and some other constraints. Namely, in our discrete 2D scenario we are given a vector field, i.e., two functions $u(x, y)$ and $v(x, y)$, to define the gradient of an image $f(x, y)$. This vector field is assumed to be arbitrary, however, in most practical scenarios, it is the result of manipulating the gradient field of some input image. Ideally, in this case we would like to find $f(x, y)$ such that $f_x(x, y) = u(x, y)$ and $f_y(x, y) = v(x, y)$. As this involves two constraints per pixel (x, y) (each yielding a single equation), while there is only one variable per pixel in $f(x, y)$, the resulting system is overconstrained, and we are not likely to find *any* function meeting these constraints. A common and natural way of overcoming this difficulty is to use a *least-squares* optimization to find a function $f(x, y)$ whose gradients are closest to $u(x, y)$ and $v(x, y)$ in the *least-squares* sense. More formally, the discrete function f

is defined by the solution of the following minimization problem:

$$\min_{f} \sum_{x,y} \left(f_x(x,y) - u(x,y)\right)^2 + \left(f_y(x,y) - v(x,y)\right)^2. \quad (7.4)$$

Differentiating this cost function with respect to all its unknowns, namely, the values of f on each pixel, and equating to zero yields the following set of equations (using the discrete definition for partial derivatives in Equation (7.3)):

$$f(x+1,y) - 2f(x,y) + f(x-1,y) + f(x,y+1) - 2f(x,y)$$
$$+ f(x,y-1) - u(x,y) + u(x-1,y) - v(x,y) + v(x,y-1) = 0.$$

Denoting $f(x+1,y) - 2f(x,y) + f(x-1,y)$ by $f_{xx}(x,y)$ and $f(x,y+1) - 2f(x,y) + f(x,y-1)$ by $f_{yy}(x,y)$, these equations read more compactly as

$$f_{xx}(x,y) + f_{yy}(x,y) = u_x(x,y) + v_y(x,y),$$

where u_x and v_y are defined analogously to f_x and f_y. Since $f_{xx}(x,y)$ and $f_{yy}(x,y)$ approximate the second derivative of f with respect to x and y, this equation can be viewed as the discrete version of the Poisson equation

$$\Delta f(x,y) = \mathrm{div}\left[u(x,y), v(x,y)\right],$$

where $\Delta f(x,y)$ denotes the discrete Laplacian $f_{xx}(x,y) + f_{yy}(x,y)$ and div is the discrete divergence operator defined by $u_x(x,y) + v_y(x,y)$. This relation also sheds light on the Laplace equation whose discrete counterpart is

$$\Delta f(x,y) = 0.$$

This equation searches for a function $f(x,y)$ with the least variation between neighboring points. Thus, these discrete analogs teach us that the Poisson equation attempts to find a function whose gradients are the closest to some given vector field, appearing on its right-hand side, and the Laplace equation looks for a function with a minimal gradient field, both in the least-squares sense. Later in this chapter, we will give a practical recipe for solving these equations.

Boundary conditions.   While examining the homogeneous Poisson and Laplace equations defined above, one quickly notices that there is no one unique solution for them. For instance, any shift of $f(x, y)$ by a constant will always be a valid solution for the Laplace equation. In order to make these solutions unique (and more controllable), we specify boundary conditions, which constrain the solution set of the equations. There are several useful types of boundary conditions, and here we will briefly describe two of them:

- The *Dirichlet* (or *first type*) *boundary condition* (named after Johann Peter Gustav Lejeune Dirichlet (1805–1859)) specifies the values that solution $f(x, y)$ needs to take on the boundary of the domain.

- The *Neumann* (or *second type*) *boundary condition* (named after Carl Neumann (1832–1925)) specifies the values that the normal derivative of the solution $\left(\frac{\partial f}{\partial \hat{\mathbf{n}}} = \langle \nabla f(x, y), \hat{\mathbf{n}} \rangle \right)$ needs to take on the boundary of the domain ($\hat{\mathbf{n}}$ is the unit vector normal to the boundary at (x, y)).

It should be noted that under some assumptions in the discrete settings, these conditions are equivalent, but we will not delve into that. It should also be noted that mixtures of these two boundary conditions exist, for instance, the *mixed boundary condition*, which employs the Dirichlet boundary condition on some parts of the boundary and the Neumann boundary condition on other parts of the boundary, or the *Robin boundary condition*, which specifies a linear combination of Dirichlet and Neumann boundary conditions (see Figure 7.2). However, we will not delve into these conditions either.

Image completion

In many situations, only a small portion of the entire data is available. In such cases, we look for means to complete the missing parts in a way that will result in a plausible completion. Having said that, the question is, what makes one solution more plausible than another? Well, in the absence of other information, a good practice is to complete the missing parts smoothly. In other words, the assumption that a minimal change in the scene occurs

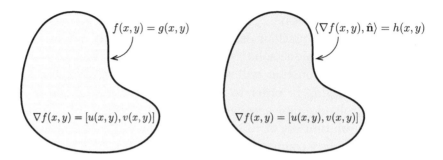

Figure 7.2: Poisson problem with Dirichlet boundary condition (left), and with Neumann boundary condition (right).

in the missing parts, which eventually leads to smooth areas, is a reasonable assumption. Thus, smoothness is a reasonable prior.

In Figure 7.3 there is a somewhat surprising result of employing the smoothness prior. The image on the left consists of several random lines. The pixels along the lines have color values taken from the corresponding pixels of the image on the right. The image on the left is completed by synthesizing a smooth image that agrees with the data along these lines, and is as smooth as possible in every other location. As can be seen, the synthesized image approximates the original image quite well, considering that the given data (in the left image) is so sparse.

Figure 7.3: Completing a sparse subset of an image (left) using the smoothness prior (middle), yields a result similar to the original image (right).

This kind of image completion is a process known as scattered data interpolation, or approximation, for which there is a rich literature (the interested reader is referred to the survey by Amidror [Ami02] and also to Chapter 10 in this book). The solution we explain here is based on the Poisson equation realized in the discrete setting of an image.

We shall start by first examining a simpler problem, where we wish to complete a single missing part (i.e., a hole) in an image (see Figure 7.4). In this problem, the pixel colors in a closed area Ω of the input image are unknown (the black hole in the lake). Instead of dealing with R, G, B values of a color image, we deal with each color channel independently, that is, we solve three independent problems. We denote by f^* the known part of the image (a scalar 2D function from (x, y) to the range of intensity values), and by f the image in the unknown area Ω.

Figure 7.4: Image completion scenario: we would like to complete the unknown 2D domain Ω in the image f^*, with the values f, such that f is continuous on the boundary of the domain $\partial\Omega$ and as smooth as possible inside that domain.

As we have already mentioned, a reasonable assumption is to complete the hole as smoothly as possible, while making sure that the values on the boundary of the hole ($\partial\Omega$) agree with the values from the original image.

In our terms, "as smooth as possible" translates to a minimum *change* in the values of f. As changes in the values of a function translate to the *derivatives* of this function, we would like the derivative of f to be as small as possible, while f on the boundary of the hole ($\partial\Omega$) should agree with f^* there. As f is a 2D function, we should consider the value of its *gradient*, as it extends the simple notion of derivative to higher dimensions. As in many other problems, we would like the sum of squares of gradients to be minimal (that is, minimizing its l_2 norm).

Formulating the above constraints yields the following minimization problem:

$$\underset{f}{\text{argmin}} \iint_{\Omega} \|\nabla f\|^2, \quad \text{s.t.} \ \ f|_{\partial\Omega} = f^*|_{\partial\Omega}. \quad (7.5)$$

At first sight, this minimization problem does not seem to help us in any way. Luckily, a field of mathematics called the *calculus of variations* leads to a simple solution of this problem. This field deals with *functionals*, as opposed to ordinary calculus, which deals with functions. Such functionals can, for example, be formed as integrals involving an unknown function and its derivatives. The interest is mainly in those functions for which the functional attains a maximal or a minimal value, which just happens to be our interest in Equation (7.5). The interested reader is referred to [GS00] for an introduction to the calculus of variations.

Specifically in our problem, the *Euler–Lagrange equation*, which is a fundamental equation in this calculus, comes into play. This equation states that if J is defined by an integral of the form $J = \int F(x, f, f_x)dx$, then J has a stationary value if the following differential equation is satisfied:

$$\frac{\partial F}{\partial f} - \frac{d}{dx}\frac{\partial F}{\partial f_x} = 0. \quad (7.6)$$

In our case, $F = \|\nabla f\|^2$, and the equation translates to

$$\frac{\partial F}{\partial f} - \frac{d}{dx}\frac{\partial F}{\partial f_x} - \frac{d}{dy}\frac{\partial F}{\partial f_y} = 0.$$

Since in our case it holds that $\frac{\partial F}{\partial f} = 0$ and $\frac{d}{dx}\frac{\partial F}{\partial f_x} = \frac{d}{dx}2f_x = 2\frac{\partial^2 f}{\partial x^2}$, Equation (7.6) reduces to

$$\frac{\partial^2 f}{\partial x^2} + \frac{\partial^2 f}{\partial y^2} = f_{xx} + f_{yy} = \Delta f = 0. \quad (7.7)$$

That is, the function that minimizes Equation (7.5), also satisfies Equation (7.7), which is the Laplace equation. As noted earlier, in order to obtain a non-trivial solution for the Laplace equation, one must also supply boundary conditions. In our case, the corresponding boundary conditions are the following Dirichlet boundary conditions:

$$f|_{\partial\Omega} = f^*|_{\partial\Omega}. \tag{7.8}$$

Now, in order to complete the image in Figure 7.4 (see Figure 7.5), all one has to do is to solve the Laplace equation with the given Dirichlet boundary conditions. A practical recipe for solving it in a discrete setting, such as the image domain, follows shortly.

Figure 7.5: The result of completing the hole in Figure 7.4 by solving the Laplace equation.

But before that, let us get back to the scattered lines problem where the missing values in the image are scattered all around the image, and not located completely inside a single domain. Again, we would like the unknown values f to smoothly interpolate the known values of f^*. As in the single domain case, we would like f to satisfy the following minimization problem:

$$\underset{f}{\operatorname{argmin}} \iint_{f^*} \|\nabla f\|^2, \text{ s.t. } f(x,y) = f^*(x,y) \text{ on the sample lines}.$$

That is, we want f to be as smooth as possible wherever its value is not a sample from f^*, and equal to f^* wherever it is a sample from it. A small difference from the minimization problem in Equation (7.5) is that now instead of integrating the gradient only inside the hole, we integrate it over the entire image domain. These equations can be easily transformed into a linear system of equations.

Now we will carry out this transformation step by step: First, for each pixel (x, y) with known values (that is, the pixel is on one of the sample lines), we can formulate the following equations which make sure that this value does not change:

$$f(x, y) = f^*(x, y), \quad (x, y) \text{ lies on one of the sample lines}. \quad (7.9)$$

For each pixel (x, y) that does not lie on one of the lines, we would like its Laplacian to be zero (aiming to minimize the gradient). That is

$$\Delta f = 0, \quad (x, y) \text{ does not lie on a sample line}. \quad (7.10)$$

Using the definition of the Laplacian operator in Equation (7.7), and the discrete approximation of partial derivatives in Equation (7.3), we get the following discrete approximation for the Laplacian operator:

$$
\begin{aligned}
\Delta f &= f_{xx} + f_{yy} \\
&= f(x+1, y) - 2f(x, y) + f(x-1, y) + f(x, y+1) \\
&\qquad\qquad\qquad\qquad\qquad - 2f(x, y) + f(x, y-1) \\
&= f(x+1, y) + f(x-1, y) + f(x, y+1) + f(x, y-1) - 4f(x, y).
\end{aligned}
$$

Thus, we can construct a system of (sparse) equations that has one equation for each pixel in the image. In order to do so, we must first introduce an alternative indexing for the image, such that we could relate to any pixel in the image by using a single index. If we denote by W, H the width and the height (respectively) of the image, in the alternative indexing, pixel (i, j) is denoted by the single index $i + jW$. We will denote the number of pixels, which is also the number of equations and the number of unknowns, by n. Note that $n = WH$.

By assuming that the known pixels are the last K equations in the set (for simplicity of display), and that the known values, respectively, are $c_1 \ldots c_k$, we can write the system in matrix notation:

$$
\begin{bmatrix}
1 & 1 & -4 & 1 & 1 & \cdots & \cdots \\
\cdots & 1 & 1 & -4 & 1 & 1 & \cdots \\
& & & \vdots & & & \\
\cdots & \cdots & 1 & 1 & -4 & 1 & 1 \\
\hline
& 0 & & | & & I_{K \times K} &
\end{bmatrix}
\begin{bmatrix}
f_1 \\
\vdots \\
f_n
\end{bmatrix}
=
\begin{bmatrix}
| \\
0 \\
| \\
\hline
c_1 \\
\vdots \\
c_k
\end{bmatrix} . \quad (7.11)
$$

This system can be solved by using a sparse linear solver, such as the ones discussed in Chapter 6.

As you have probably noticed, the linear equation system we have described above has no notion of the fact that the missing values all lie on scattered lines. That is, the same system also solves the image completion problem described in Equation (7.5).

Image operations by Poisson equations

After we saw the usage of the Laplace equation for image completion, in this section we will examine several more applications of the Laplace and Poisson equations in the image domain.

Poisson image editing. In many situations, we would like to copy some characteristics of an image to another image, in a smooth manner. An example for such a situation is the seamless cloning operator [PGB03], which seamlessly copies a source image patch into a target image. In such a cloning process, we want to incorporate the source patch into the target patch such that no apparent discontinuities occur, while preserving the original nature of the source patch. The basic idea behind *Poisson image editing* is guiding the Poisson-based completion process to fill the missing parts of an image using gradients from another source image. Figure 7.6 shows two examples of Poisson image cloning, taken from [JSTS06] and [FHL⁺09]. A free-form selection containing a wooden log (left image) is copied and pasted onto two beaches (two right columns). For each target beach, the middle image is the result of simply replacing the pixels with the pixels from the source image, while the bottom image is the result of copying the gradients of the source image patch and reconstructing the image by a function f whose gradients agree with the gradients of the pasted patch, but at the same time also agree with the boundaries of the patch. Note how the colors of the wooden log and its surroundings have adapted so that the cloned patch agrees with the target image.

Formally, we are endeavoring to solve the following minimization problem:

$$\operatorname*{argmin}_{f} \iint\limits_{\Omega} \|\nabla f - \mathbf{g}\|^2, \quad \text{s.t. } f|_{\partial\Omega} = f^*|_{\partial\Omega}. \qquad (7.12)$$

Figure 7.6: Poisson image cloning. A wooden log (left) is cloned into two beaches. The upper row contains the original beaches, the center two images are the result of simply copying the pixels, and the result of applying Poisson image cloning is shown in the bottom row.

Again, f is the new image we construct, Ω is the region where f is reconstructed, f^* is the target image, and $\mathbf{g} = [g_1, g_2]$ is a guidance map used to guide the image editing to achieve a desired goal. A common value for $\mathbf{g}$, when performing seamless cloning is ∇f_{source}, where f_{source} is the source image.

We should note that although this minimization problem is similar to the one in Equation (7.5), it is not identical, as now the integrand is $\|\nabla f - \mathbf{g}\|^2$ instead of simply $\|\nabla f\|^2$. This time the corresponding Euler–Lagrange equation reduces to the Poisson equation and is

$$\frac{\partial^2 f}{\partial x^2} + \frac{\partial^2 f}{\partial y^2} = \frac{\partial g_1}{\partial x} + \frac{\partial g_2}{\partial y} \, .$$

Often, the above equation is written concisely as

$$\Delta f = \operatorname{div} \mathbf{g},$$

where

$$\operatorname{div} \mathbf{g} = \frac{\partial g_1}{\partial x} + \frac{\partial g_2}{\partial y}.$$

which can be discretized by

$$\operatorname{div} \mathbf{g} = \frac{\partial g_1}{\partial x} + \frac{\partial g_2}{\partial y}$$
$$= g_1(x, y) - g_1(x-1, y) + g_2(x, y) - g_2(x, y-1).$$

As before, this equation can be linearized and plugged into a linear system, the only difference now is that the right-hand side contains the values of $\operatorname{div} \mathbf{g}$ instead of zeros. Again, for the simplicity of display, the last K equations correspond to the K known values on the boundary. The corresponding system is:

$$
\begin{bmatrix}
1 & 1 & -4 & 1 & 1 & \cdots & \cdots \\
\cdots & 1 & 1 & -4 & 1 & 1 & \cdots \\
& & & \vdots & & & \\
\cdots & \cdots & 1 & 1 & -4 & 1 & 1 \\
\hline
& 0 & & | & & I_{K \times K} &
\end{bmatrix}
\begin{bmatrix}
f_1 \\
\vdots \\
f_n
\end{bmatrix}
=
\begin{bmatrix}
| \\
\operatorname{div} \mathbf{g} \\
| \\
\hline
c_1 \\
\vdots \\
c_k
\end{bmatrix}.
$$

$$(7.13)$$

Healing brush. A common task in an everyday image editing process is the concealment of areas in a photo. This concealment might be of a complete object in a scene, see Figure 7.7, or of very

Figure 7.7: Concealment of the Caracol Falls.

local artifacts or unwanted details, as in Figure 7.8. The basic mechanism behind this process is heavily based on Poisson equations. Namely, very similar to Poisson image cloning, by importing seamlessly a piece of the background, complete objects, parts of objects and undesirable artifacts can easily be removed. Usually, multiple user strokes are used to mark up the regions that need to be concealed.

Figure 7.8: Removing the spots from a cheetah.

Panoramic stitching.   Another example of gradient domain image editing is stitching large panoramic images (see Figure 7.9). Nowadays, high-resolution (10 mega-pixel and more) cameras are easily available, which allows amateur photographers to take large panoramic images. In the problem of *image stitching*, we are given a set of input images with overlapping areas, and the goal is to stitch these images into a large coherent image. In general, we can split this problem into three parts, as described below.

In the first part, all the images are registered, such that identical features in different images are aligned against each other, and placed in the same location in the output image. As the registration process is beyond the scope of this chapter, the interested reader is referred to [SS97] for an example of an image registration operator, and to [Low99] for a classic example of the SIFT descriptor, which is commonly used for feature detection.

In the second part, seams are defined between adjacent image patches, such that each pixel in the output image originates directly from a pixel in a specific source image. This is usually achieved using a graph cut-based algorithm (see Chapter 12 for more details).

Figure 7.9: Panoramic image stitching. The result when simply compositing the three images of the panorama (visible seams highlighted) (top) and the result when applying Poisson panoramic stitching (bottom). Images are from [Aga07].

In the third and last part, we must eliminate artifacts remaining in the region along the seams. These artifacts stem from the fact that the lighting conditions and camera parameters may have changed between the two adjacent images. This part is usually carried out in the gradient domain.

Laplace on meshes

The Laplace equation (as defined in Equation (7.1)) is dimensionless, and it can be applied over various domains. In the previous sections we have shown its use over the image space domain. The image space is a very convenient 2D grid, where the Laplace operator (Equation (7.7)) is easy to compute. It is also possible to define the Laplace operator over an irregular grid, although the partial derivatives are not as trivially defined as in the regular grid case. In the regular image space grid, the degree of each node is four. In an irregular grid, the degree is not fixed and can vary between three and eight or more. The generalization of the discrete Laplace operator to irregular meshes was discussed in Chapter 5. Let us recall here that in the simplest form, a discrete mesh Laplacian of a vertex $\mathbf{f} \in \mathbb{R}^3$, denoted by $\Delta(\mathbf{f})$, is

$$\Delta(\mathbf{f}) = \mathbf{f} - \frac{1}{d}\sum_{k=1}^{d} \mathbf{f}_k \ ,$$

where d is the degree (number of neighbors) of the node $\mathbf{f}$, and $\mathbf{f}_k$ are neighbors of vertex $\mathbf{f}$.

The above equation enables generalizing the principle expressed in Equation (7.7) to irregular meshes. Setting the Laplacian $\Delta(\mathbf{f})$ of a given mesh to zero while imposing some boundary conditions leads to the construction of a minimal surface that passes through some preset locations. The surface will look like a membrane and have spikes around the fixed locations. On the other hand, *minimizing the norm* of $\Delta(\mathbf{f})$ of a given mesh under positional constraints, as follows:

$$\min_{\mathbf{f}} \int \|\Delta(\mathbf{f})\|^2, \ s.t. \ \mathbf{f}_j = \mathbf{c}_j \ \forall j \in \mathcal{C},$$

leads to a smooth surface (also at the fixed locations), which is more useful in geometry processing. Here, the $\mathbf{c}_j$'s are the present vertex locations, and $\mathcal{C}$ is the set of indices of fixed vertices. The Euler–Lagrange equations for the above minimization amount to the *bi-Laplace equation*:

$$\Delta^2(\mathbf{f}) = 0, \ s.t. \ \text{positional constraints are fulfilled.}$$

The bi-Laplace operator Δ^2 amounts to composing the Laplace operator with itself (i.e., applying the Δ operator twice). For more details on this equation, the reader is referred to [BS08].

Figure 7.10 shows the camel mesh reconstructed by solving the bi-Laplace equation over the given mesh with a number of fixed, preset vertices as boundary conditions. In the close-up of the reconstructed head, we can see the preset vertices in red. On the left, the mesh is reconstructed with 100 preset vertices, while on the right there are 200 known vertices. Solving the bi-Laplace equation reconstructs an approximation of a thin-plate surface which respects the preset vertices. For more details, the interested reader is referred to [SCO04].

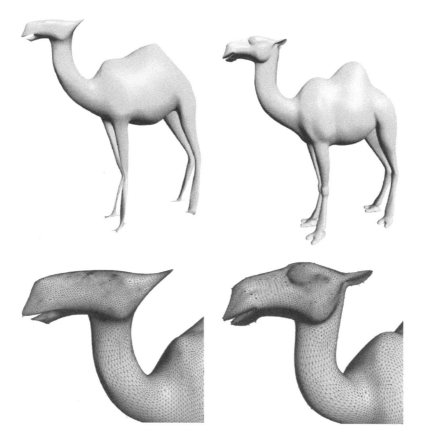

Figure 7.10: A camel mesh reconstructed by solving the bi-Laplace equation, with 100 (left) and 200 (right) preset vertices, respectively.

Acknowledgments. The *Baba* image in Figure 7.3 was taken by Flickr user Wen-Yan King and is used under the Creative Commons Attribution 2.0 License.

The portraits of Laplace and Poisson in Figure 7.1 are in the public domain as their copyrights have expired.

We would like to thank the following Flickr users for the permission to use their imagery: Mr. Tiago Fioreze (Figure 7.7) and Mr. Mark Probst (Figure 7.8).

Chapter 8

Curvatures: A Differential Geometry Tool

Niloy J. Mitra and Daniel Cohen-Or

The term *curvature* simply refers to the amount of bending of a curve or a surface. In simple terms, curvature at a point indicates how quickly a curve or a surface bends away from its local tangent line or tangent plane, respectively. While tangent planes are useful for studying first-order behavior of surfaces, curvatures deal with second-order behavior. However, a rigorous study of curvatures can be quite involved, and in general, there are different notions of curvatures depending on the context. The interested reader is referred to the comprehensive survey of curvature in the classical differential geometry text by Do Carmo [dC76]. In this chapter, we will explain the basic ideas, develop some intuitions, and study the necessary formulations that allow us to better understand the notion of curvature and help us to compute various types of curvatures.

Partial surface matching

Before studying curvatures, let us first understand the relevance to partial surface matching, which is a core component of geometry processing. Let us take a look at the *Buddha* model in Figure 8.1. The four lotus flowers that are part of the model are

Figure 8.1: Using a curvature-based local surface descriptor, it is possible to extract parts of an object that are self-similar to other parts of the same object. This is an instance of the partial matching problem. A careful local analysis coupled with geometric hashing, enables the extraction of geometrically similar flowers from the *Buddha* model [GCO06].

highlighted. The flowers were automatically detected by a self-similarity algorithm [GCO06]. To detect such similarity among portions of a larger geometric model requires a *partial matching* mechanism.

Matching is a fundamental task in a vast number of geometric applications in numerous fields such as computer vision, robotics, and molecular biology [VH99]. Let us focus on the problem of partial matching between 3D models given by a surface representation rather than as volume. Here we are interested in partial matching, which is a much harder problem than global matching, where the similarity is measured between entire models [TV04]. Partial matching is complicated by the fact that we simultaneously need to search for and define the sub-parts prior to measuring similarities. That is, the parts that are matched are not predefined, can be any sub-shape of a larger shape, and in any orientation or scale. Partial matching can occur between any regions across the given surfaces, leading to an expensive combinatorial problem [MPWC12].

One way to cope with the combinatorial complexity is to reduce the number of potential candidates for partial matching. One can carefully select a relatively small set of points across the surfaces, and use surface patches around them as candidates for the similarity test. Such an approach for partial matching requires an efficient mechanism to match similar surface patches. The difficult part is to associate each surface patch with a local shape descriptor or, say, a *geometric signature*, which well represents the surface patch. These signatures can then be fed into a hashing mechanism to quickly retrieve a small set of similar candidates. While there are a lot of studies on hashing, here we are merely focusing on the design of the geometric signature. An efficient signature should be compact so that many of them can be stored in the hash table, and it should be effective in the sense that it well approximates the geometric similarity among the original surface patches that it represents. There are many ways to define a signature; one popular and effective signature is known as the *spin-image* [Joh97]. However, here we will turn to a simple signature that is directly based on a fundamental concept from differential geometry—surface curvature.

Curvature as a local surface descriptor

A fundamental property of a curvature is that it is a differential surface property that is not affected by the choice of the coordinate system, or the parameterization of the surface. In particular, curvature is a rigid invariant embedding in space. A *rigid invariant property* is an essential property of a geometric signature that stays unchanged under relative position and orientation changes. It is an essential property of a geometric signature to allow partial matching between similar geometric surface patches independent of their position and orientation in space, which can be arbitrary.

Differential analysis and invariance properties.[1] As stated previously, curvature is invariant to rigid transformations. More formally, it is an *extrinsic* differential property: it characterizes the shape of the surface in the *embedding* space. Imagine a non-elastic sheet of paper: its *extrinsic* properties characterize how it

[1]Can be left out in the first reading.

is folded. The differential analysis toolbox also provides *intrinsic* quantities that measure properties of the space *embedded* on the surface, e.g., a 2D domain for 3D surfaces. Consider a sheet of paper. A 10-centimeter line drawn on it will always measure 10 centimeter irrespective of how the sheet is folded. In other words, one can compute the distance between two points on a surface by measuring their *extrinsic* Euclidean distance, e.g., the shortest path in the *embedding* space, or their *intrinsic* geodesic distance, e.g., the shortest path in the *embedded* space. As a result, *intrinsic* quantities are invariant to isometries, i.e., transformations that do not change the surface metric, and *extrinsic* properties are invariant to rigid transformations, a subclass of isometries. A complete study of *intrinsic* properties is beyond the scope of this chapter.

Observe that the effect of the rigid transformation of the patches can be nullified by representing the surface patch, for example, by relative coordinates with respect to the patch center and its normal. In Figure 8.2, we observe how a local coordinate frame can be encoded in terms of curve tangent and normals, and how it moves along with the curve under rigid transforms. Hence, such a local coordinate system remains invariant under rigid transformations. Recently, Mitra et al. [MGP06] used curvatures as local surface descriptors for detecting partial symmetries in 3D geometry.

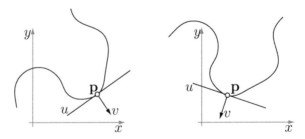

Figure 8.2: A local coordinate frame at any point **p** of a curve can be defined in terms of local tangent and normal lines. Such a coordinate frame, *uv*-frame in this example, being attached to the curve, moves along with the curve under any rigid transformations. Hence, curve attributes defined in terms of such a local coordinate system remain *invariant* under rigid transformations.

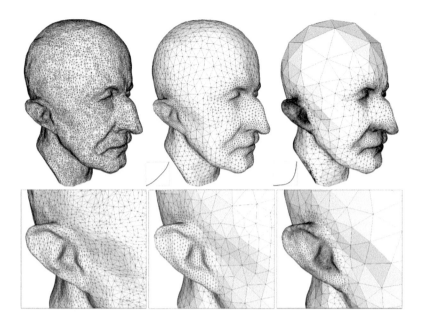

Figure 8.3: Curvature-sensitive remeshing [AMD02]. Since curvature measures how fast the surface moves away from its local tangent plane, or in other words, how curved the surface is, better surface approximation is achieved by using smaller triangles in regions of high curvature.

Surface meshing and simplification

In geometric modeling, we are often required to represent a model given a particular polygon budget. Such compact models lead to faster model transmission, rendering, and manipulations. Given a highly detailed model, we simplify the same by compressing the detailed regions to meet our polygon budget. In order to identify the detailed regions, curvature is a natural choice since it gives us a local measure of how curved the surface is. Thus, curvatures come up naturally in attribute preserving simplification.

Curvature helps us decide where a model has to be represented in detail: Flat regions, or regions of low curvature, can be well represented by only a few polygons, while for curved regions, or regions of high curvature, we proportionally need more polygons to accurately represent the underlying surface (see Figure 8.3). In Figure 8.4, we see application of curvature lines for extracting an efficient polygonal representation [ACSD+03].

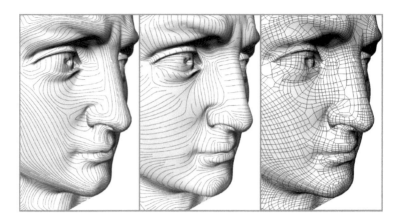

Figure 8.4: Using principal curvatures (shown in blue and red), a surface can be efficiently represented in terms of polygons. Such a curvature-aware polygonal representation of the surface, as shown in the right figure, helps to better retain geometric features, thus preserving surface qualities using a given polygon count [ACSD+03].

Curvature of a curve

Curvatures measure the extent of bending. This is a differential property that can be defined as a measure of the rate of deviation between the surface and the tangent plane of that surface at a given point.

Let us first examine the notion of curvature in 1D, where a surface reduces to a curve in the xy plane. The curvature of a point $\mathbf{p}$ along the curve is evaluated as the second derivative at $\mathbf{p}$ (see Figure 8.5). To compute that, let us build a local "uv" frame, where the origin is at $\mathbf{p}$ and the u axis coincides with the tangent direction at $\mathbf{p}$, and v is orthogonal to u, i.e., the local normal direction at $\mathbf{p}$. Then the curvature at $\mathbf{p}$ is simply the second derivative computed in this uv-frame measuring how fast the local normal changes, or in other words how fast the surface bends. Locally, the best second-order approximation to the curve is given by $v = \kappa u^2/2$, where κ denotes the local curvature at $\mathbf{p}$. Remember that tangent planes are locally the best linear approximations. Note that our choice of coordinate frame is the local tangent normal; hence, in the second-order approximation, we do not have any linear term. The behavior is similar for surfaces,

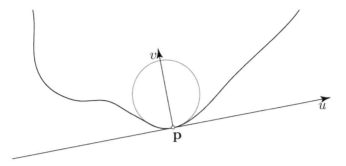

Figure 8.5: At a point $\mathbf{p}$, a curve is to be locally expressed in its local tangent-normal coordinate frame. At point $\mathbf{p}$, the best second-order approximation is related to its *osculating* circle: the circle that best touches or kisses the given curve at $\mathbf{p}$. If the radius of the circle is r, curvature is defined as $\kappa(\mathbf{p}) = 1/r$. Locally, the surface can be represented as $(u, \kappa u^2/2)$.

where a local second-order surface can be defined in terms of the two principal curvatures.

Curvature can also be directly computed in the original xy coordinate system: Assuming the curve is a function $f(x)$ in the xy plane, then the curvature $\kappa(x)$ at point x is given by

$$\kappa(x) = \frac{f''(x)}{(1 + f'(x)^2)^{3/2}} \, . \tag{8.1}$$

The nominator simply corrects the second derivative to compensate for the fact that the second derivative is not necessarily computed over a local frame that agrees with the x direction. However, if the x axis is tangent to the function at x, then $f'(x) = 0$ and the above equation reduces to $\kappa(x) = f''(x)$.

Note that the curvature is a rigid invariant property as it ignores the constant (translation) and the first derivative (rotation). It only measures the speed by which the tangential direction (the first derivative) changes along the curve, or in other words, computing the second derivative. The third- and higher-order derivatives have no effect on the second derivative at the origin. Factoring out the rigid motion is achieved by constructing the local frame at the point in question, and the curvature is actually the speed by which this local frame rotates when we move infinitesimally along the curve.

If a curve is specified in a parametric form, $x = x(t)$ and $y = y(t)$, then the curvature at t can be directly computed using

$$\kappa(t) = \frac{x_t y_{tt} - y_t x_{tt}}{(x_t^2 + y_t^2)^{3/2}}, \tag{8.2}$$

where $x_t = dx/dt$, $x_{tt} = d^2x/dt^2$, $y_t = dy/dt$, $y_{tt} = d^2y/dt^2$, respectively.

Curvature of surfaces

Let us now extend this definition of curvature to a two-dimensional surface. Analogously, at any given point on the surface we construct an orthogonal local uvw coordinate system such that the uv plane is tangent to the surface and the w axis is perpendicular to the surface at that point, i.e., w denotes the local surface normal. Curvature measures the speed by which this local frame changes. However, now we have to be careful since the local frame has additional degrees of freedom compared to the curve case. When the local frame moves along a given curve over the surface, it may rotate with two degrees of freedom, in contrast to one degree of freedom in case of curves. We can measure curvature as the speed by which the normal component is changing, ignoring the torsion component.

Locally, the surface expressed in the uvw coordinate system defines a function over a uv plane, and is known as a *Monge patch*. Dealing with a height function simplifies many computations. For example, now we can easily compute the curvature along any tangent *direction* **j**. The curvature can be measured along a curve defined by the intersection of the surface with a vertical plane, given by the normal plane, along the **j** direction as studied in the previous section (see Figure 8.6). This curvature again measures how the normal to the curve rotates as we move along the curve. This is also known as the *normal curvature* at a surface point along the direction **j**.

However, we are not interested in the curvature in only a particular direction, but along all directions emanating from a point. Interestingly, it is sufficient to study curvatures only along two

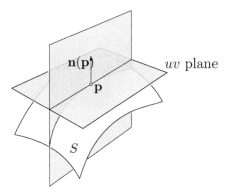

Figure 8.6: Locally, the surface S is a function above a uv plane, the tangent plane at $\mathbf{p}$, and the curvature of a curve can be computed along a given direction. The curve is defined by the intersection of a plane orthogonal to uv and containing the given direction and the normal vector $\mathbf{n}(\mathbf{p})$.

special directions, the directions where the corresponding normal curvatures get their extremum values. These directions are known as the *principal curvature directions*, and the corresponding curvature values, denoted by $\kappa_{\max}$ and $\kappa_{\min}$, are the maximum and minimum curvature values. Further, since these principal directions can be shown to be the eigenvectors of a special matrix, the principal directions are always mutually orthogonal.

The beauty of the these principal curvatures and directions is that they capture a lot of information about the local surface. Again, curvatures on a surface deal with second derivatives, ignoring the higher-order behavior of the surface. Analogous to the curve case, surface curvatures give us a local quadratic patch approximation of the surface (see Figure 8.7).

Now let us see how can we compute and find these principal curvatures and directions. We know how to compute the derivatives along the u and v directions, simply by taking derivatives along the respective directions. There is a little magic in the local Hessian matrix of $w(u, v)$, also known as the second fundamental form of the local quadratic patch, which consists of easy-to-compute partial derivatives along the u and v directions:

$$H(u, v) = \begin{bmatrix} w_{uu} & w_{uv} \\ w_{uv} & w_{vv} \end{bmatrix}. \tag{8.3}$$

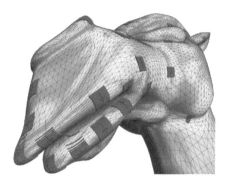

Figure 8.7: Local quadric surfaces, marked in red, computed at a number of points on the camel model. The crosses indicate the principal curvature directions.

The principal curvatures and directions are the eigenvalues and eigenvectors, respectively, of the $H(u, v)$. Note that the Hessian matrix is a 2×2 matrix, for which there are simple closed forms to compute the eigenvectors and eigenvalues ($\kappa_{\max}, \kappa_{\min}$).

Let us now look at a couple of simple examples:

1. In the case of a plane, it is easy to see that the eigenvalues of the corresponding Hessian are both zeros, resulting in the conclusion that the principal curvatures for a plane are both zero. This is quite intuitive since we know that a plane is *flat*.

2. For a sphere of radius r, the Hessian at any point on the sphere can be shown to be $\begin{bmatrix} 1/r & 0 \\ 0 & 1/r \end{bmatrix}$. Hence both the principal curvature values on any point of a sphere are equal to $1/r$, i.e., the reciprocal of the sphere's radius.

Surface analysis

Having seen how to compute principal curvatures, let us use it to understand local surface behavior. Based on their signs, we can categorize surfaces into six basic classes as shown in Table 8.1. The actual values of the principal curvatures contain much more information, i.e., the degree of convexity or concavity of the local patch.

	$\kappa_{\min} < 0$	$\kappa_{\min} = 0$	$\kappa_{\min} > 0$
$\kappa_{\max} < 0$	peak	ridge	saddle
$\kappa_{\max} = 0$	ridge	flat	valley
$\kappa_{\max} > 0$	saddle	valley	pit

Table 8.1: Locally categorizing surfaces based on the principal curvatures signs.

The two principal curvatures can be combined into a single number, known as the *shape index* [KvD92], which is defined by

$$S = -\frac{2}{\pi} \arctan \left(\frac{\kappa_{\max} - \kappa_{\min}}{\kappa_{\max} + \kappa_{\min}} \right). \qquad (8.4)$$

With the shape index, each local shape is associated with a value in the range of $[-1, 1]$, where all convex shapes have positive shape indices and all concave shapes have negative shape indices. Saddle points are mapped close to the zero value, where the minimal saddle point is assigned to zero. Note that when $\kappa_{\max} = \kappa_{\min}$ there are no *unique* principal directions, and the sign of the curvatures define it to be either purely convex or concave. Such points, where the principal curvature values are equal, are called *umbilical* points.

The shape index, or the two principal curvatures, form a simple rigid-invariant shape signature that can be used to efficiently compare different surface patches. Dissimilar signatures can be used to quickly prune potential matches, while needing to test for matching only between surface patches that have similar shape index signatures.

Gaussian and mean curvatures

The term *curvature* in the context of surface is ambiguous since it can refer to multiple things. Typically, either the *Gaussian* or *mean* curvature are used in the context of surfaces. Previously we have learned about *principal* curvatures. While the principal curvatures are associated with the principal directions, the Gaussian and mean curvatures are directionless quantities.

While Gaussian and mean curvatures can be introduced in multiple ways, the following is a simple one based on the principal curvatures: The Gaussian curvature K of the surface is the product of the principal curvatures:

$$K = \kappa_{\max}\kappa_{\min}, \qquad (8.5)$$

and mean curvature H of the surface is the average of the principal curvatures:

$$H = \frac{\kappa_{\max} + \kappa_{\min}}{2}. \qquad (8.6)$$

It immediately follows that when K and H are known, the principal curvatures are the roots of the quadratic equation:

$$\kappa_{\max}, \kappa_{\min} = H \pm \sqrt{H^2 - K}. \qquad (8.7)$$

It should now be clear that the Gaussian and mean curvatures are rigid-invariant quantities. Further, it can be shown that Gaussian curvature is invariant under isometry, and is thus an *intrinsic* surface property. For example, whenever a surface is bent, without introducing kinks (like gently bending a sheet of paper), geodesic distances (surface metric), angles or area remain unchanged. But when a sheet of rubber is deformed, the Gaussian curvature changes. If the Gaussian curvature $K = 0$ everywhere on the surface (like a sheet of paper, for example), the surface is called a *developable surface*. On the other hand, the mean curvature changes even when the surface bends isometrically. Surfaces with vanishing mean curvature almost everywhere are known as *minimal surfaces*.

Gaussian and mean curvatures can be computed explicitly if Monge patch representation of the surface is available. Given a Monge patch of the form $w(u, v)$, the Gaussian curvature at a point $(u, v, w(u, v))$ can be directly computed as

$$K(u, v) = \frac{w_{uu}w_{vv} - w_{uv}^2}{(1 + w_u^2 + w_v^2)^2}, \qquad (8.8)$$

while the corresponding mean curvature is given by:

$$H(u, v) = \frac{(1 + w_v^2)w_{uu} - 2w_u w_v w_{uv} + (1 + w_u^2)w_{vv}}{2(1 + w_u^2 + w_v^2)^{3/2}}. \qquad (8.9)$$

Subsequently, we can compute the principal curvature values using Equation (8.5).

Building a Monge patch. Having studied curvatures for surfaces, and how to compute them given Monge patches, we are left with the task of building Monge patches given a surface. In other words, given a point with surface normal and the corresponding tangent plane, how do we locally approximate the surface using a height field with respect to the tangent plane? Luckily, this is something we have already learned: Recall how we constructed a local least-squares (LS) surface approximation in Chapter 3. Around a point $\mathbf{p}$, if the $\mathbf{u}, \mathbf{v}$ denote a set of orthogonal directions on the local tangent plane, then using LS fitting we can approximate the surface in terms of $(u, v, w(u, v))$ thus constructing a Monge patch. Since we are interested in measuring curvatures, a quadratic LS fit is sufficient.

Gauss–Bonnet theorem.[2] In differential geometry, we study the local behavior of curves and surfaces. Surprisingly, even though everything is local and deals with differential properties, there is a strong global coupling ensured by global topological invariants; the Gauss-Bonnet theorem makes this relation precise. We look at a simpler version of this general theorem.

We start with some definitions. *Total curvature* of a region refers to the integral of Gaussian curvature on the enclosed region. Given two points on a surface, a *geodesic* refers to a shortest path connecting the two points such that the path stays on the surface. Now, given three surface points, we construct a geodesic triangle by connecting each pair of points by a geodesic path (see Figure 8.8). The total curvature of such a geodesic triangle equals the deviation of the sum of its angles from π, i.e.,

$$\sum_{i=1}^{3} \theta_i = \pi + \int_R K dA, \qquad (8.10)$$

the geodesic triangle region being denoted by R.

Let us look at a simple example of a sphere of radius r (Figure 8.8). Let the origin be the center of the sphere, and the three axes pass through $\mathbf{p}$, $\mathbf{q}$, and $\mathbf{r}$, respectively. The Gaussian curvature at any point on the sphere is given by $1/r^2$. Putting things

---

[2]Can be left out in the first reading.

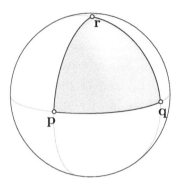

Figure 8.8: Geodesic triangle Δ**pqr** on a sphere. The integral of Gaussian curvature in the enclosed region can be computed using a special relation between the angles of the triangles and the integral of Gaussian curvature. This is a special case of the Gauss–Bonnet theorem, a famous theorem that relates differential geometry to topology.

together we have,

$$
\begin{aligned}
\sum_{i=1}^{3} \theta_i &= \pi + \int_{\Delta \mathbf{pqr}} 1/r^2 dA \\
&= \pi + (1/r^2)(\Delta \mathbf{pqr}) \\
&= \pi + (1/r^2)(4\pi r^2/8) \\
&= 3\pi/2 \,.
\end{aligned}
$$

Since, all the three angles are symmetric and equal, we conclude that each angle is $\pi/2$.

Such global properties can be used to check validity of local computations, and also to increase precision of numerical computations. However, for our current purpose, it is enough to be aware that such a relation exists between differential geometry and topology.

Concluding remarks

In this chapter, we have studied about curvatures for curves and surfaces. Curvatures are fundamentally coupled with object geometry, and hence come up often as compact object descriptors. The interested reader should learn about the first and second fundamental forms [dC76], and ideally take a full course in differential geometry.

Chapter 9

Dimensionality Reduction

Hao (Richard) Zhang and Daniel Cohen-Or

In this chapter, we learn the concept, usefulness, and execution of dimensionality reduction. This important topic has been extensively studied in different disciplines. A large variety of techniques have been developed and there have been numerous articles in the literature including several accessible surveys [BDR+06, HLMS04] and even books [CC94, STC04]. Here we shall present and discuss only a sampler of techniques and illuminate them using visually intuitive examples.

Generally speaking, dimensionality reduction seeks to reduce the dimensionality of a given data set, mapping high-dimensional data into a lower-dimensional space to facilitate visualization, processing, or inference. The mapping is typically done in an information preserving manner: depending on the problem at hand, the mapping can retain the most relevant information in the data and presents it in the more manageable low-dimensional space. Here the term *manageable* can be interpreted in different ways. A typical example in the field of visualization maps high-dimensional data into two or three dimensions that are a lot more intuitive to us when displayed. In other applications, carrying out tasks in a low-dimensional space brings focus to the essence of the problem at hand so that the analysis becomes more straightforward. As well, dimensionality reduction obviously speeds up computation since the data size is reduced.

Even though the original input data may reside in a high-dimensional space, effective reduction is possible in many instances since the true structure of the data, or at least the structure that we care about in an application, is inherently low-dimensional. An example of the former is a flat piece of paper rolled and bent in however complex ways; while the resulting shape resides in three-dimensional space, it is still intrinsically two-dimensional. Dimensionality reduction thus seeks to reveal the intrinsic or the most relevant structure of given data. When information loss is inevitable, the goal is to minimize that loss when dimensionality is reduced. Our coverage starts with linear dimensionality reduction, which is typically executed by principal component analysis (PCA). The more interesting case of non-linear dimensionality reduction and associated applications are then presented.

Linear dimensionality reduction

Those readers who are familiar with the notion of data representations using a set of basis vectors will appreciate that linear dimensionality reduction techniques are mainly about finding a compact set of bases to best approximate useful data. Linearity stems from the fact that the data vectors are expressed as a linear combination of the basis vectors. The technique of PCA, which we covered in Chapter 4, is most frequently used in these situations due to their optimal approximation or compaction properties [Jai89].

Eigenfaces for recognition. A classic example of PCA-based linear dimensionality reduction is the use of *eigenfaces* for face image recognition [TP91]. Given a database of face images capturing different individuals, the recognition problem seeks the identity of a query face within the database. For simplicity, let the database be formed by N grayscale face images. We assume that the faces are well aligned in the images and were captured under similar lighting conditions. Each image has the same size $n \times m$ (the total number of pixels) and presents itself as a face descriptor or signature $\mathbf{u}_i$, a vector consisting of nm pixel intensity values, $i = 1, \ldots, N$. Thus, a face corresponds to a point in an nm-

dimensional space. Even for a moderate choice of $n = m = 64$, the face signatures are of very high dimension and lead to high computational cost when comparing the different faces for recognition.

To obtain more compact face signatures, a linear transform can be applied to the $\mathbf{u}_i$'s to reduce their dimensionality via PCA. Specifically, we form the $nm \times nm$ covariance matrix Σ for the face data set $\{\mathbf{u}_1, \ldots, \mathbf{u}_N\}$,

$$\Sigma = \sum_{i=1}^{n}(\mathbf{u}_i - \overline{\mathbf{u}}) \cdot (\mathbf{u}_i - \overline{\mathbf{u}})^\top \text{ with } \overline{\mathbf{u}} = \frac{1}{n}\sum_{i=1}^{n}\mathbf{u}_i.$$

To obtain Σ, the data are first *centered* by subtracting the mean $\overline{\mathbf{u}}$ from them. The (i, j)-th entry of Σ expresses the covariance, which is a dot product, between the data representations in the i-th and j-th dimensions.

The eigenvectors of Σ, referred to as the eigenfaces, span the face set. The principle eigenvectors or eigenfaces, those corresponding to leading eigenvalues of Σ, represent the *major modes* of variations of the face images about the average face. Figure 9.1 shows several eigenfaces obtained for the well-known Yale face database. As we can see, each eigenface accentuates certain fa-

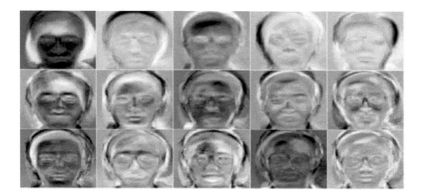

Figure 9.1: The first 15 eigenfaces obtained for the Yale face image database (http://www.cs.princeton.edu/~cdecoro/eigenfaces/). Describing each face image by its pixel intensities results in a high-dimensional (e.g., 64 × 64) feature vector. Alternatively, each face can be well approximated by linear combinations of a small number of (e.g., 40) eigenfaces.

cial characteristics. Computationally, one does not need to solve
the eigendecomposition problem on Σ, which is of size $nm \times nm$.
Typically, N, the number of faces in the dataset, is much smaller
than $nm \cdot nm = n^2m^2$, thus Σ does not have full rank (its rank
is N). It follows that to obtain the eigenfaces, it suffices to solve
only an eigendecomposition problem of size $N \times N$; see the last
section of Chapter 4 for an explanation of this "trick."

Typically, a rather small number of principal eigenfaces, e.g.,
$K = 40$, are sufficient to provide good approximations to the faces
in the database. In other words, when reconstructing each face
signature $\mathbf{u}_i$ as a linear combination of the K eigenfaces, the reconstruction error is small. In fact, the theory of PCA ensures that
for any given K, reconstruction using the K leading eigenfaces is
optimal in terms of residual errors measured using the squared
Euclidean norm. It follows that the *face space*, the space of reasonable faces including all those from the database, is expected
to be well characterized by the 40 or so eigenfaces through their
linear combinations. In that sense, the face space is essentially
low-dimensional. For face recognition, each face image can be effectively and compactly represented using a description consisting
of K coefficients, the weights in the linear combination using the
K eigenfaces. This represents a dimensionality reduction from
nm, e.g., 4,096 for $n = m = 64$, to about 40, greatly improving
the efficiency of storage and face retrieval in the database.

Morphable models for face synthesis. The formulation of
eigenfaces can be easily extended to 3D face data as long as a
point-to-point correspondence is established between all the 3D
face representations. Here the vector $\mathbf{u}_i$ consists of the x, y, and
z coordinates of the vertices of a 3D face mesh. Beyond recognition tasks, the face space characterized by the 3D eigenfaces can
also be utilized for face synthesis, as in the work of Blanz and
Vetter [BV99] on morphable models. In this work, a new face
is always produced as a linear combination of existing faces in
the database. For efficiency, all faces are represented using their
eigenface coefficients.

To only synthesize faces that are reasonable, or shall we say
"human-like," they should not deviate too much from the exam-

ple faces. Thus, the choice of the new eigenface coefficients is regulated by a probabilistic distribution. Specifically, each coefficient is controlled by the variation of the faces in the database along the corresponding direction given by the eigenface; these variations are given by the eigenvalues of the covariance matrix in the PCA formulation. Indeed, from the theory of PCA, while the eigenvectors of the covariance matrix provide the modes or axes of variations of the data, the eigenvalues are related to the magnitude of the variations; see Chapter 4 for detailed coverage.

Non-linear methods

PCA is a fundamental dimensionality reduction technique, which works well as long the space, like the space of face images, behaves very much like a linear space. Characteristics such as the ability to classify entities using *linear separators* (hyperplanes in high-dimensional space) and having modes of variations and direction of variations along *straight axes* reflect such linearity. However, as we shall see next, the intrinsic structure of the data may very well be non-linear, which calls for non-linear methods; some of these methods are essentially non-linear versions of PCA.

Figure 9.2: A thumbnail summarizing a motion sequence.

Our first motivating example involves visualizing motion sequences, a problem studied by Assa et al. [ACCO05]: given a motion sequence, express it in a concise yet characteristic still image. The key idea is to analyze the complex motion sequence and extract a handful of *key poses* that can then be extrapolated to form a still figure to serve as an overview, such as the one shown

in Figure 9.2. To appreciate the problem's difficulty, we first need
to understand how motion sequences are represented.

Figure 9.3: A series of skeletal poses in a motion sequence.

An animation or motion sequence typically consists of a series
of skeletal poses, as illustrated in Figure 9.3. The skeleton is defined by a number of bones connected by joints and the poses are
parameterized by spatial or physical attributes including positions,
angles, and velocities associated with the bones and joints. Rich
and realistic motions necessitate a large number of such attributes.
Thus, the animation traces out a *high-dimensional motion curve*,
one that is embedded in the space parameterized by possibly hundreds of attributes describing the skeletal poses.

The high dimensionality is an obstacle since applying any analysis directly to a high-dimensional curve is inefficient—the "curse
of dimensionality." However, embedding the curve in a low-
dimensional space allows a rather simple geometric algorithm to
be applied to identify important poses. In their work, Assa et
al. [ACCO05] embedded the motion curves in five to seven dimensions. Moreover, to facilitate visualization of the curves, an
embedding to the three-dimensional space can be constructed, as
shown in Figure 9.4(a). Such a familiar curve visualization, along
with the marked key poses, conveniently reveals that these poses
tend to occur at geometric features along the curve.

Data analysis, exploration and visualization are fundamental
tasks in many areas of science. When faced with a large amount of
multivariate data, dimensionality reduction can be essential. The
key challenge is to find a compact and expressive representation of
high-dimensional data in a lower-dimension space so as to reveal
the true structure of the data. To appreciate the need for non-

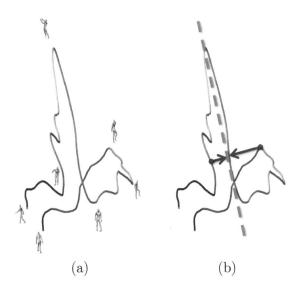

(a)   (b)

Figure 9.4: An illustration of a series of skeletal poses along a motion sequence depicted by a curve. The curve is mapped from a high-dimensional space into the 2D plane to facilitate its visualization. (a) Several key poses are shown, and it is evident that they typically correspond to geometric features along the curve. (b) Linear PCA properly reflects the curve structure. In particular, two points far apart along the motion sequence get mapped into the same point along the first principal PCA axis.

linear methods, we note that applying the classical PCA to points along the motion curve in Figure 9.4 does not yield useful results. Looking at the first principal component, shown as the blue line in Figure 9.4(b), we see that it obviously does not capture the distance relationship between points along the non-linear motion curve. If the true structure of a set of points is merely a 1D curve embedded in a high-dimensional space, then the right dimensionality reduction scheme should *unfold* that curve into a straight line—the intrinsic dimension of the curve is one-dimensional. However, it is not always so obvious what that intrinsic dimension or true structure is, as exemplified by our next example.

Our second example is a classical one from clustering where the input data is shown in Figure 9.5(a). The most natural clustering should return the three rings of points, yet the three rings cannot be linearly separated in the original space of the input points,

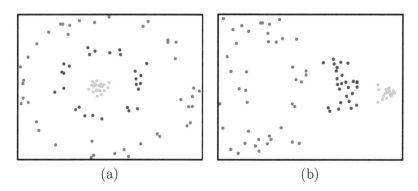

Figure 9.5: (a) A point set roughly depicting three concentric rings and (b) its embedding using kernel PCA, where the kernel is defined as $K(\mathbf{x}, \mathbf{y}) = (\mathbf{x}^\top \mathbf{y} + 1)^2$. It is clear that a linear separator along the first principal component, the horizontal axis in (b), allows us to cluster the data correctly (image reproduced from Wikipedia page on kernel PCA).

the 2D plane. It is not so obvious how dimensionality reduction is relevant in the three-ring example as the points already reside in a low-dimensional space, the 2D plane. The key to note here is that we are interested in the *clustering structure* of the data, yet the right separators for the clusters are obviously non-linear— they are circular in this case. To allow linear separators to do the job, we must map the data points to a higher-dimensional space, often referred to as a *feature space*. This should not come as a surprise since non-linear phenomena always require more parameters (forming a higher-dimensional parameter space) to express compared to their linear counterparts.

To be effective, we would like the mapping to feature space to allow linear methods to work. The desire to work with linear methods is due to their simplicity and well-developed theory and algorithms with examples including the classical PCA and clustering or classification methods such as k-means and support vector machines. Finally, dimensionality reduction comes in since the dimensionality of the feature space might be too high and the true dimensionality of the data, after having been mapped to the feature space, is much lower.

To be convinced that high dimensions would help, consider that we are allowed to map a set of n points into an n-dimensional

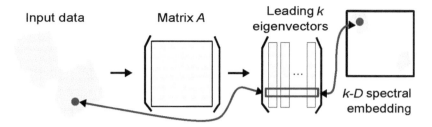

Figure 9.6: The overall approach of non-linear methods using spectral embedding. The operator or matrix A encodes some pairwise relationships among the input points. The spectral embedding is obtained from a set of leading eigenvectors of A.

space. Specifically, the i-th point is mapped to the n-dimensional point P_i whose i-th coordinate is 1 and all other coordinates are zero. It can be verified that any two subsets of $\{P_1, \ldots, P_n\}$ can be linearly separated by a hyperplane. The non-linear dimensionality techniques discussed in this chapter all share the common characteristic that they map the input data into a high-dimensional feature space and then reduce the dimensionality to finally map the data into a lower-dimensional *embedding* space to execute the task as hand.

Algorithmically, the common approach is to express all pairwise relations, e.g., distances, similarities, or other forms of affinities, between the input data points in a linear operator or matrix A and then compute an embedding. Typically, the embedding is derived from only a few leading eigenvectors of A since they capture the characteristics of A sufficiently well; see Figure 9.6 for an illustration. Due to the use of eigenvectors in this setting, the embeddings are commonly called *spectral embeddings*.

The differences between the different types of non-linear dimensionality reduction techniques mainly lie in how the data affinities (matrix A) are defined. We reiterate that although the dimensionality of the original input data may be high, the purpose of these methods is not to reduce that dimensionality, but to reduce the high dimensions inherent from the non-linearity of the relations associated with the input data for the task at hand such as clustering or visualization.

Kernel PCA

When thinking of a mapping Φ into a high-dimensional feature space, an obvious question is whether it is tractable to perform operations in that space, which may be infinite-dimensional. The now well-known *kernel trick* (yes, it is an official name) corresponds to a mechanism through which a non-linear problem can be solved by performing linear analysis in the feature space without ever explicitly computing the mapping Φ. The kernel trick is applicable to any algorithm that can be formulated in terms of dot products exclusively, as such the mapping Φ is never explicitly computed and it is only known and applied through the dot products $\langle \Phi(\mathbf{x}), \Phi(\mathbf{y}) \rangle$. PCA is one algorithm that fits the above criterion. As mentioned earlier, PCA performs eigendecomposition on the covariance matrix, which is constructed by dot products. It can be shown that PCA in the feature space can be equivalently carried out by using the eigenvectors of a kernel matrix [SSM97]; this is referred to as *kernel PCA*, the topic of discussion in this section. With kernel PCA, the mapping Φ can be as high-dimensional as possible so as to unfold the data sufficiently for the analysis task at hand. In practice, the mapping Φ is only an abstraction and not known or expressed explicitly; we simply work with the dot products.

Let $\mathbf{x}_1, \ldots, \mathbf{x}_n \in \mathbb{R}^N$ be the input points and their feature space mappings be $\Phi(\mathbf{x}_1), \ldots, \Phi(\mathbf{x}_n) \in \mathbb{R}^M$; often $M \gg N$. To perform PCA, we assume that $\Phi(\mathbf{x}_1), \ldots, \Phi(\mathbf{x}_n)$ are centered, i.e., they have zero mean. Here we recall again that PCA operates on centered data. Then, given any point $\mathbf{x}$ in the input space, the projection of its feature-space image $\Phi(\mathbf{x})$ onto the k-th principal component $\mathbf{v}_k \in \mathbb{R}^M$ derived from PCA using the $\Phi(\mathbf{x}_i)$'s can be computed as

$$\langle \mathbf{v}_k, \Phi(\mathbf{x}) \rangle = \mathbf{e}_k^\top \mathbf{g}, \tag{9.1}$$

where $\mathbf{e}_k$ is the k-th eigenvector of the matrix $K \in \mathbb{R}^{n \times n}$ defined by

$$K_{ij} = \langle \Phi(\mathbf{x}_i), \Phi(\mathbf{x}_j) \rangle, \ i, j = 1, \ldots, n, \tag{9.2}$$

and the entries of the vector $\mathbf{g} \in \mathbb{R}^n$ are given by $\mathbf{g}_i = \langle \Phi(\mathbf{x}_i), \Phi(\mathbf{x}) \rangle$, $i = 1, \ldots, n$. Note that the left-hand side of Equation (9.1) is the dot product of two high-dimensional vectors in the feature space, while the right-hand side is the dot product of two size-n vectors. We shall not provide a proof of the above claim here; interested readers should refer to [SSM97].

From Equation (9.1), we see that to derive the embedding of point $\mathbf{x}_j$, $1 \leq j \leq n$, via kernel PCA, we simply let $\mathbf{x} := \mathbf{x}_i$. We are thus projecting the i-th row of the matrix K onto $\mathbf{e}_k$. Since $\mathbf{e}_k$ is the k-th eigenvector of K, $K\mathbf{e}_k = \lambda_k \mathbf{e}_k$, the resulting projection is nothing but the j-th entry of $\mathbf{e}_k$ scaled by the k-th eigenvalue λ_k of K. If the projection is performed on the first k eigenvectors, we obtain a k-dimensional spectral embedding of the point.

In general, we cannot assume that the feature-space images $\Phi(\mathbf{x}_1)$, ..., $\Phi(\mathbf{x}_n)$ are centered, i.e., they have zero mean. To center them, the matrix that we are to eigendecompose becomes

$$\overline{K} = K - \mathbf{1}_n K - K \mathbf{1}_n - \mathbf{1}_n K - \mathbf{1}_n, \qquad (9.3)$$

where $\mathbf{1}_n$ is the $n \times n$ matrix with $(\mathbf{1}_n)_{i,j} = 1/n$.

A keen observer would note that so far we have been careful not to call the matrix K a kernel matrix. Indeed, the analysis above is applicable to any K that takes the form in Equation (9.2). In our context, a kernel matrix is one that is positive semi-definite. There is a theorem from functional analysis, called Mercer's theorem, which says that if K is a kernel matrix, then its entries can be expressed as dot products in a high-dimensional space as given by Equation (9.2).

When applying kernel PCA in practice, it is crucial that we choose an appropriate kernel matrix K. K should be positive semi-definite. Even though we never compute the mappings to feature space, through K, we are able to impose constraints on the mapped points defining their dot products. In Figure 9.5, we show one result of kernel PCA on the three-rings dataset. As we can see in (b), in the embedding space defined by the first two principal components, the three rings can be linearly separated, reflecting the effectiveness of kernel PCA. As we mentioned before, from an application point of view, the embedding allows simpler (linear) clustering or classification techniques to be applied, making kernel PCA one of the most valuable tools for these applications. In this example and generally for others, the kernel matrix models some form of affinity between the input points, e.g., a Gaussian of the Euclidean distance. Hence the matrix K is often referred to as an *affinity* matrix. In kernel PCA, the affinities are to be reproduced by the feature-space images through their dot products.

Multidimensional scaling

Multidimensional scaling, or MDS, is a set of related techniques often employed in data visualization [CC94]. In classical MDS, low-dimensional embeddings, typically 2D, are constructed to facilitate visualization of high-dimensional data. Given a set of n input points $\mathbf{x}_1, \ldots, \mathbf{x}_n$ and some $n \times n$ distance matrix M defined by the user, $M_{i,j} = \text{dist}(\mathbf{x}_i, \mathbf{x}_j)$, $i, j = 1, \ldots, n$, the goal of MDS is to produce a set of embeddings of the input points whose pairwise *Euclidean* distances in the embedding space provide a good approximation to the given distances in M. Note that the distance or pairwise dissimilarity measure $dist(\mathbf{x}_i, \mathbf{x}_j)$ is not required to be a metric; it can be as general as any user-defined distance.

To perform MDS, we first apply *double centering* and normalization to M, akin to Equation (9.3). This results in the matrix

$$B = -\frac{1}{2} JMJ,$$

where

$$J = I - \frac{1}{n} \mathbf{1}\mathbf{1}^\top$$

and $\mathbf{1}$ is the column vector of 1's. Then we compute the eigenvalues and eigenvectors of B. The spectral embeddings of the input points are derived from the eigenvectors, which are scaled by the square roots of the corresponding eigenvalues; see Figure 9.6 for an illustration, but note an extra step of scaling the eigenvectors. The nice thing about this particular spectral embedding, as illuminated by a theorem proved by Eckart and Young [EY36], is that the Euclidean distances between the spectral embeddings closely approximate the distance values in M in the sense of the Frobenius norm, the usual error metric for matrices.

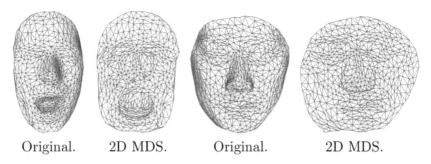

| Original. | 2D MDS. | Original. | 2D MDS. |

Figure 9.7: 2D multi-dimensional scaling (MDS) applied to two face meshes flattens the meshes while trying to preserve the mesh geometry in the plane.

In addition to data visualization, MDS can also be applied to flatten a surface mesh embedded in 3D space, e.g., for the purpose of texture mapping [ZKK02]. That is, the surface mesh is embedded in the plane in order to obtain texture coordinates for its vertices and the 2D coordinates of the mesh vertices are computed by MDS. Since one of the primary goals of texture mapping is to reduce distortion, that is, discrepancies between distances within the texture map and distances over the mesh surface, the distance matrix M used in MDS is given by the pairwise *geodesic* distances (shortest distances over the mesh surface) between the mesh vertices. In Figure 9.7, we show two examples of 2D MDS embeddings of face meshes.

Pose normalization

Since the non-linear dimensionality reduction techniques we have covered so far all operate solely on the affinity matrix, different input data that imply the same or similar pairwise affinities will possess the same or similar spectral embeddings. This observation leads to an effective means of handling object pose in the analysis of articulated shapes such as humans and other creatures.

Two important applications here are shape retrieval and correspondence, both crucial tasks in digital content creation. In retrieval, one wishes to extract shapes from a data repository that are similar to a given query shape. Obviously, we would like the similarity measure to be invariant to the pose of the relevant objects. Shape correspondence is needed when one wishes to transfer pre-defined motions on one character to another. Again, pose-invariance is sought when corresponding two character shapes.

Under different poses of an articulated shape, the shape bends at the joints but incurs little stretching; in reality, a small amount of stretching does occur near the joints. Therefore, the geodesic distances over the shape surface remain largely unchanged. It follows that defining affinities based on geodesic distances can effectively remove the influence of pose from shape analysis.

Elad and Kimmel [EK03] use MDS based on geodesic distances to compute bending-invariant shape signatures. The resulting spectral embedding effectively normalizes the mesh shapes with respect to translation, rotation, and bending transformations. The similarity between two shapes is then given by the Euclidean distance between the moments of the first few eigenvectors, usually less than 15, where moments are certain global and scalar descriptions of one-dimensional functions or vectors. These similarity distances between moments can be used for shape classification.

Jain and Zhang [JZvK07] rely on higher-dimensional embeddings based on the eigenvectors of an affinity matrix to obtain point correspondence between two mesh shapes. The affinities are defined by using geodesic distances between points over the shape surfaces, which are approximately invariant to pose changes. The first k eigenvectors of the affinity matrix are used to embed the model in a k-dimensional space; typically $k = 5$ or 6. Figure 9.8 shows 3D embeddings of a few articulated shapes for visualization purposes, where we can observe pose normalization. After the process is performed on two models, the two embeddings are aligned and the correspondence between the two shapes is given by the proximity (Euclidean distances) between the vertices after such alignment. Evidently, pose normalization via spectral embedding greatly facilitates the shape alignment.

More recently, Au et al. [ACOT$^+$10] perform spectral embedding on skeletal representations of shapes for pose normalization; see Figure 9.9. Pairwise geodesic distances between skeletal nodes (marked by circles in the figure) are used to define a distance ma-

Figure 9.8: 3D spectral embeddings (bottom row) of some articulated 3D shapes (top row) from the McGill 3D shape benchmark database. Since the mesh operator is constructed from geodesic distances, the embeddings are normalized with respect to shape bending.

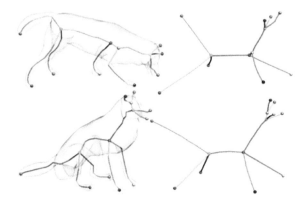

Figure 9.9: 3D spectral embeddings (right) of skeletal representations of two dog models (left); note the pose normalization.

trix. The embeddings are obtained via multi-dimensional scaling. By factoring out pose, the correspondence between shape skeletons is facilitated. In turn, correspondence between the shapes themselves is derived from skeleton correspondence.

Chapter 10

Scattered-Data Interpolation

Tao Ju

A classical problem in mathematical and geometric modeling is to obtain a continuous function from a small number of samples. The problem arises in various applications. In 3D scanning, for example, the scanner collects the locations of sample points on some unknown continuous surface, which needs to be determined from these samples.

To introduce and define the problem, let us consider a 2D example in Figure 10.1: given a few gray-colored dots on the left, how can we compute a complete image, such as the one on the right, that interpolates the intensity at each dot? Note that the problem is ill-posed, because there could be infinitely many such

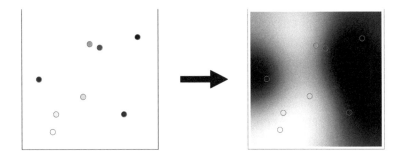

Figure 10.1: Interpolating gray dots (left) with a gray image (right).

images that interpolate. In most scenarios, the image is expected to exhibit a *nice* look where the criteria of *niceness* depends on the particular application.

More generally, the interpolation problem can be stated as follows: given a set of N-dimensional *data points* $\{\mathbf{p}_1, \ldots, \mathbf{p}_n\}$ associated with scalar *data values* $\{v_1, \ldots, v_n\}$, compute a function $f(\mathbf{p}) : \mathbb{R}^N \to \mathbb{R}$ for any point $\mathbf{p}$ in the N-dimensional space such that $f(\mathbf{p}_i) = v_i$ for $i = 1, \ldots, n$. In the example above, the data points reside on a 2D plane ($N = 2$), and the data values are intensities at each point. We are particularly interested in the case where the data points are *scattered*, or distributed with a non-uniform density, just like those dots in Figure 10.1. Such data is quite common in practice because a uniform sampling is not always available.

To further motivate our discussion, we will look at a well-known application of scattered-data interpolation in geometric modeling—surface reconstruction.

Example application: surface reconstruction. There are a number of ways to capture a real-world 3D object as a cloud of points on the object's surface. One may use a 3D scanning device, or apply image-based reconstruction algorithms to digital pictures. The point cloud that is obtained using these methods usually lacks connectivity information. It is desirable for a number of purposes, such as visualization, analysis and simulation, to *connect* these scattered points to form a complete surface. For ease of discussion, let us first examine the 2D example in Figure 10.2(a), where the point cloud is highlighted by blue circles and needs to be connected to form a closed curve.

To formulate the reconstruction task as an interpolation problem, we could consider each point $\mathbf{p}_i$ in the cloud as a data point associated with some intensity value, say 0. If we can obtain an interpolating function $f(\mathbf{p}) : \mathbb{R}^2 \to \mathbb{R}$ so that $f(\mathbf{p}_i) = 0$ for every $\mathbf{p}_i$, the reconstructed curve would comprise all points $\mathbf{p}$ (including the data points $\mathbf{p}_i$) where $f(\mathbf{p}) = 0$. Unfortunately, this formulation has a trivial, uninteresting solution: the constant function $f(\mathbf{p}) \equiv 0$, in which case the curve becomes the entire plane.

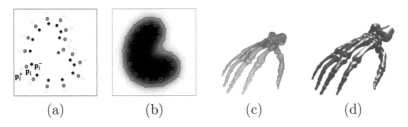

Figure 10.2: Reconstructing (b) a closed 2D curve or (d) a 3D surface from point clouds (highlighted in blue in (a,c)). Pictures (c,d) are from [CBC+01].

To get around the trivial solution, we need to constrain the interpolation problem with more data points that have nonzero intensities. A common approach is to place artificial data points with positive or negative intensities at locations inside or outside the curve to be reconstructed. To find such locations, an outward unit normal vector $\hat{\mathbf{n}}_i$ is first estimated at each point $\mathbf{p}_i$ in the original point cloud, and new points are placed at $\mathbf{p}_i^+ = \mathbf{p}_i + d\hat{\mathbf{n}}_i$ and $\mathbf{p}_i^- = \mathbf{p}_i - d\hat{\mathbf{n}}_i$ (highlighted by green and red circles in Figure 10.2(a)) where d is a small positive scalar. These points will be associated with intensities proportional to d or $-d$. The additional constraints enforce the interpolating function $f(\mathbf{p})$ to become negative inside the curve and positive outside, as plotted in Figure 10.2(b), where $f(\mathbf{p}) = 0$ is a closed curve connecting the original point cloud (colored blue in Figure 10.2(b)).

The interpolation-based approach can be easily extended to 3D. Once we have an interpolating function $f(\mathbf{p}) : \mathbb{R}^3 \to \mathbb{R}$, the surface $f(\mathbf{p}) = 0$ can be reconstructed using grid-based contouring methods such as the Marching Cubes. An example of the reconstruction result is shown in Figure 10.2(c,d), taken from the work of [CBC+01].

There are many methods that one may consider for computing functions that interpolate. In this chapter, we will get to know an important member of this family, known as *radial basis functions* (RBFs). One of the nice features of RBFs is that they result in smooth functions with low oscillation, as seen in both Figures 10.1 and 10.2. This property makes the RBF an ideal choice in many applications including surface reconstruction from point clouds.

Rather than introducing RBFs immediately, we will look at a succession of interpolation methods whose methodologies bare strong similarity to RBFs. These methods, namely *piecewise linear interpolation* and *Shepard's interpolation*, are simpler and also commonly used in applications. But they yield functions that are less smooth or more oscillating than those generated by RBFs. For example, Figure 10.3 (top) shows the interpolated image given the input in Figure 10.1 using the three methods. The differences are better visualized in Figure 10.3 (bottom) where each interpolating function $f(\mathbf{p})$ is plotted as a height surface. That is, the height at a 2D point $\mathbf{p}$ is $f(\mathbf{p})$. Our hope is that by first visiting the two simpler methods, it will become easier to motivate and understand RBFs.

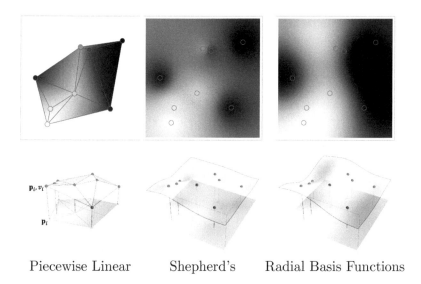

Piecewise Linear   Shepherd's   Radial Basis Functions

Figure 10.3: Comparing different interpolation solutions for the gray dots in Figure 10.1 (left), showing the 2D gray image (top) and the 3D function terrain (bottom).

Piecewise linear interpolation

For ease of discussion, we will start with a one-dimensional setting. Here, the data points $\mathbf{p}_i$ lie on a single coordinate axis. The simplest input consisting of only two data points is shown in Figure 10.4(a), where each pair of data point and value $\{\mathbf{p}_i, v_i\}$ is plotted as a 2D point. Intuitively, a function $f(\mathbf{p})$ that interpolates the values at these two data points would form a curve that passes through the two 2D points.

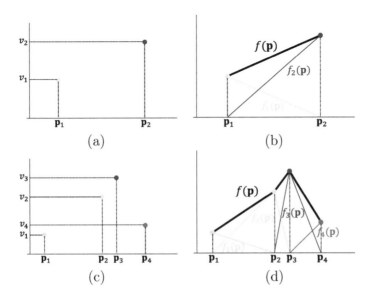

Figure 10.4: Linear interpolation between two points (top) and multiple points (bottom) in 1D.

An obvious solution in this case is to connect the 2D points with a line segment, such as the black line in Figure 10.4(b). The function corresponding to the underlying line of this segment has the form:

$$f(\mathbf{p}) = \frac{\mathbf{p}_2 - \mathbf{p}}{\mathbf{p}_2 - \mathbf{p}_1} v_1 + \frac{\mathbf{p} - \mathbf{p}_1}{\mathbf{p}_2 - \mathbf{p}_1} v_2 . \tag{10.1}$$

This formula is known as *linear interpolation*. It is easy to verify that $f(\mathbf{p}_i) = v_i$ for $i = 1, 2$. For reasons that will soon become apparent, we will limit such interpolation to be between the two data points, that is, for $\mathbf{p} \in [\mathbf{p}_1, \mathbf{p}_2]$.

Before moving on to multiple data points, we give an alternative interpretation of linear interpolation, which will be used in the rest of the chapter to motivate other approaches. Observe that the right-hand side of equation (10.1) is a sum of two terms, each being a linear function of $\mathbf{p}$. We denote these two functions as $f_1(\mathbf{p})$ and $f_2(\mathbf{p})$, and plot them in color in Figure 10.4(b). The sum of these two functions, which are line segments themselves, is the final interpolating segment.

To interpolate between more than two data points, we can linearly interpolate between every two successive points, and connect the resulting line segments into a polyline (this is why we consider interpolation only between two data points in equation (10.1)). Following the alternative interpretation of linear interpolation, the polyline $f(\mathbf{p})$ can be constructed as the sum of a set of polylines $f_i(\mathbf{p})$, one for each data point $\mathbf{p}_i$, and has the piecewise form

$$
f_i(\mathbf{p}) = \begin{cases} \frac{\mathbf{p}-\mathbf{p}_{i-1}}{\mathbf{p}_i-\mathbf{p}_{i-1}}v_i, & \text{if } i > 1 \text{ and } \mathbf{p} \in [\mathbf{p}_{i-1}, \mathbf{p}_i), \\ \frac{\mathbf{p}_{i+1}-\mathbf{p}}{\mathbf{p}_{i+1}-\mathbf{p}_i}v_i, & \text{if } i < n \text{ and } \mathbf{p} \in [\mathbf{p}_i, \mathbf{p}_{i+1}], \\ 0, & \text{otherwise}, \end{cases} \qquad (10.2)
$$

for $\mathbf{p} \in [\mathbf{p}_1, \mathbf{p}_n]$.

An example with 4 data points is shown in the bottom of Figure 10.4. Observe that each polyline $f_i(\mathbf{p})$ has a bump-like shape. We will call them (and later definitions of $f_i(\mathbf{p})$) *bumps* or *bump functions*. To verify that the sum of these bumps, $f(\mathbf{p}) = \sum_{i=1}^n f_i(\mathbf{p})$, interpolates the data, we observe that each bump satisfies $f_i(\mathbf{p}_j) = \delta_{ij}v_i$ for $i, j \in [1, n]$, where the delta function δ_{ij} is defined as 1 if $i = j$ and 0 otherwise. As a result, $f(\mathbf{p}_i) = v_i$ for $i = 1, \ldots, n$.

To apply piecewise linear interpolation to data points in higher dimensions, we will first need to tessellate the data points into elementary domains, within which linear interpolation is performed. In the two-dimensional example of Figure 10.3 (left), the gray dots are first connected to form triangles (a process called *tessellation*), and within each triangle, the gray values at the vertices of the triangle are linearly interpolated.

As seen in the 1D and 2D examples, piecewise linear interpolation results in continuous functions. However, the derivatives of these functions $f(\mathbf{p})$ are not continuous (e.g., at the data points

$\mathbf{p} = \mathbf{p}_i$), resulting in jagged or faceted appearance. In addition, in two dimensions and beyond, the interpolation also depends on how the data points are tessellated, which is not unique in general. For example, Figure 10.5 shows two different ways in which the data points from Figure 10.1 can be tessellated, resulting in different interpolations. To get a smoother interpolation, the two approaches we look at next will employ smoother bump functions than that in equation (10.2). In addition, the definition of these bump functions will not require a tessellation of the data points.

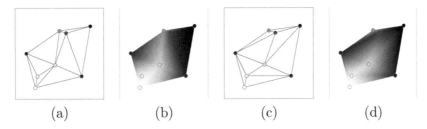

(a)   (b)   (c)   (d)

Figure 10.5: Two ways (a,c) of tessellating the data points in Figure 10.1 yield different piecewise linear interpolation results (b,d).

Shepard's interpolation

One of the earliest approaches for tessellation-free interpolation of scattered data was named after Shepard [She68] and later improved by others [FN80]. Similar to piecewise linear interpolation, Shepard's interpolation can also be expressed as a sum of bump functions. The value of each bump function $f_i(\mathbf{p})$ in Shepard's interpolation decays from v_i to zero as $\mathbf{p}$ moves away from the "center" of the bump, $\mathbf{p}_i$:

$$f_i(\mathbf{p}) = \frac{(\|\mathbf{p} - \mathbf{p}_i\|^{-\alpha})}{\sum_{j=1}^{n} (\|\mathbf{p} - \mathbf{p}_j\|^{-\alpha})} v_i \,, \qquad (10.3)$$

where $\| \cdot \|$ is the Euclidean distance metric, and the exponent α is a positive number. To verify that the sum of the bumps $f(\mathbf{p}) = \sum_{i=1}^{n} f_i(\mathbf{p})$ interpolates the data, observe that the fraction in the equation approaches 1 as $\mathbf{p}$ approaches $\mathbf{p}_i$ (which causes both the numerator and the denominator to go to infinity) and 0 as $\mathbf{p}$ approaches $\mathbf{p}_j$ (which causes the denominator to go to infinity, but not the numerator).

The bump function $f_i(\mathbf{p})$ defined in Equation (10.3) is once-differentiable for $\alpha > 1$, and so is their sum $f(\mathbf{p})$. An example of Shepard's interpolation in 1D is shown in Figure 10.6 for $\alpha = 2$, using the same set of data points in Figure 10.4 (bottom). Observe that the Shepard's bumps have a smoother shape than those used in piecewise linear interpolation. In addition, Shepard's interpolation is not restricted to the region confined by the data points, but extends out (or *extrapolates*) to the entire domain.

The same formulation in Equation (10.3) can be directly used to interpolate data points in higher dimensions, without need for tessellation. Figure 10.3 (middle) shows the result of Shepard's interpolation of the set of 2D gray dots in Figure 10.1, yielding a smooth gray image. Again, note that the image is defined everywhere on the plane due to the extrapolatory nature of Shepard's interpolation.

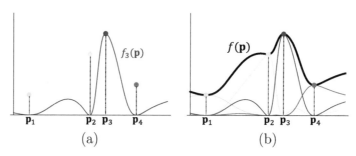

Figure 10.6: Shepard's interpolation in 1D: (a) a bump function at $\mathbf{p}_3$, (b) the interpolating function (black curve) as the sum of all bumps.

Although Shepard's interpolation results in a smooth function, this function has a rather undesirable property that it "flattens out" at data points, as can be observed in both 1D and 2D examples. To understand why, observe from Figure 10.6(a) that each bump function $f_i(\mathbf{p})$ has zero derivatives at $\mathbf{p} = \mathbf{p}_j$ for all $j = 1, \ldots, n$. As a result, the sum of the bump functions, $f(\mathbf{p})$, also has zero derivatives at those locations.

The flat tops in Shepard's interpolation introduce extra "waviness" into the interpolating function, and makes the function look unnecessarily more complex than the data itself. Ideally, if the data points and values are sampled from a simple shape (such as a straight line), we would like the interpolating function to reproduce this shape exactly. In the next section, we will see how a less wavy interpolation can be achieved using smooth bump functions with few places that have zero derivatives. The approach also allows function reproduction by slightly modifying the bump-adding strategy.

Radial basis functions

There are many bump-shaped functions that are smooth and have few places with zero derivatives. For example, the Gaussian function,

$$\phi(r) = e^{-\frac{r^2}{2c^2}}, \tag{10.4}$$

for $r \geq 0$ is a 1D bell-like curve with height 1. The function has zero derivative only at $r = 0$. The parameter c controls the width of the bump. The full width of the bump at half the maximum height (0.5) equals $2\sqrt{2 \ln 2}\, c$.

We can construct a bump function $f_i(\mathbf{p})$ centered at data point $\mathbf{p}_i$ using the Gaussian function by applying it to the Euclidean distance between $\mathbf{p}$ and $\mathbf{p}_i$,

$$f_i(\mathbf{p}) = \phi(\|\mathbf{p} - \mathbf{p}_i\|)v_i. \tag{10.5}$$

A 1D example of this bump (which we will refer to as a *Gaussian bump*) is shown in Figure 10.7(a). Similar to the Shepard's bump (Figure 10.6(a)), the Gaussian bump peaks at $\mathbf{p} = \mathbf{p}_i$ with height v_i and goes to zero as $\mathbf{p}$ moves away from $\mathbf{p}_i$. Compared to Shepard's bump, the Gaussian bump has a much less wavy shape.

Unfortunately, simply taking the sum of the Gaussian bumps does not give an interpolating function, as shown in Figure 10.7(b). To understand why, observe that a Gaussian bump $f_i(\mathbf{p})$ is non-zero everywhere, including at $\mathbf{p} = \mathbf{p}_j$ for $j \neq i$, unlike the bumps we used in Shepard's or piecewise linear interpolation. To be able to interpolate while maintaining the nice shape of the Gaussian

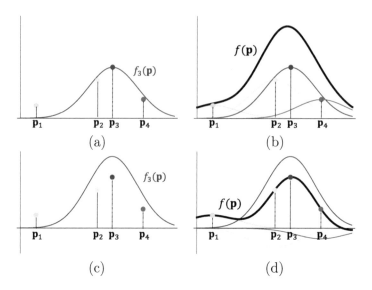

Figure 10.7: Comparing the sum of Gaussian bumps (top) with the sum of radial basis functions (bottom): the latter interpolates while the former does not.

bumps, we make a small adjustment to Equation (10.5),

$$f_i(\mathbf{p}) = \phi(\|\mathbf{p} - \mathbf{p}_i\|)c_i\,, \qquad (10.6)$$

where v_i in the Gaussian bump are replaced by coefficients c_i. Intuitively, c_i are free variables that allow us to adjust the height of a Gaussian bump, so that the sum of these bumps interpolates values at all data points. This bump function is known as a *radial basis function* (RBF).

Our goal then is to find the RBF coefficients c_i such that $f(\mathbf{p}_i) = \sum_{j=1}^{n} f_j(\mathbf{p}_i) = v_i$ for $i = 1, \ldots, n$. These constraints result in the following linear system,

$$A\,\mathbf{c} = \mathbf{v}\,,$$

where

$$A = \begin{bmatrix} \phi(\|\mathbf{p}_1 - \mathbf{p}_1\|) & \cdots & \phi(\|\mathbf{p}_1 - \mathbf{p}_n\|) \\ \vdots & \ddots & \vdots \\ \phi(\|\mathbf{p}_n - \mathbf{p}_1\|) & \cdots & \phi(\|\mathbf{p}_n - \mathbf{p}_n\|) \end{bmatrix},$$

$$\mathbf{c} = [c_1, \ldots, c_n]^\top,$$

$$\mathbf{v} = [v_1, \ldots, v_n]^\top.$$

$$(10.7)$$

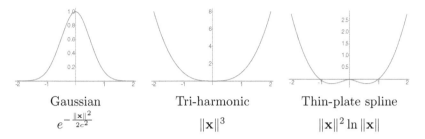

Figure 10.8: Radial functions.

Since A is a square matrix, the unknowns $\mathbf{c}$ can be solved using a standard numerical solver.

A one-dimensional example of RBF interpolation is shown in Figure 10.7 (bottom). Compared to Shepard's interpolation (Figure 10.6), the RBF yields functions interpolating the same input data but that are not as wavy. This advantage is amplified in the 2D example in Figure 10.3, where the RBF interpolation result is shown on the right.

There are many other radial function candidates ϕ that can be used in place of the Gaussian in Equation (10.6). Figure 10.8 shows two other commonly used functions, the thin plate spline, $\phi(r) = r^2 \ln r$, and the tri-harmonic function, $\phi(r) = r^3$, for $r \geq 0$. Observe that both convex or concave bumps can be used, since the coefficients c_i can be negative. In some radial functions (such as tri-harmonic), we have $\phi(0) = 0$, but this is also fine since the interpolating function is the sum of all radial functions. An example of interpolation using the tri-harmonic functions is shown in Figure 10.9. The use of any of these functions ensures that the matrix A in Equation (10.7) has full rank and is thus invertible.

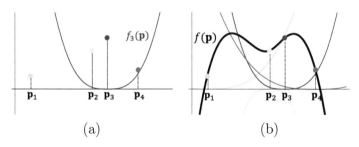

Figure 10.9: RBF interpolation using tri-harmonic radial functions.

Reproducing polynomials. We mentioned in the discussion of Shepard's interpolation that it would be desirable for an interpolation function to be no more complex than the original function from which the data points and values are sampled. As an evaluation, we show in the top row of Figure 10.10 several 1D examples where the data points and values are sampled from low-degree polynomials, including a constant function (left), a linear function (middle), and a quadratic function (right). The second row shows RBF interpolation of these data. Unfortunately, the interpolating functions still appear more complex and wavy than the original functions.

We can enrich the RBF to exactly reproduce polynomials up to a given degree. To do so, we introduce a polynomial term into the interpolation function in addition to the sum of all bump functions,

$$f(\mathbf{p}) = g(\mathbf{p}) + \sum_{i=1}^{n} f_i(\mathbf{p}), \qquad (10.8)$$

where $g(\mathbf{p})$ is a polynomial of maximum degree k. In essence, we will represent any polynomial components in the data (with degree k or less) using $g(\mathbf{p})$ and represent the residue using the bumps $f_i(\mathbf{p})$. The polynomial $g(\mathbf{p})$ will be solved together with the RBF coefficients so that $f(\mathbf{p})$ interpolates.

To solve for the k-degree polynomial $g(\mathbf{p})$, we first need a general representation of any k-degree polynomial. This can be done by a weighted sum of monomials,

$$g(\mathbf{p}) = \sum_{i=1}^{l} a_i \, m_i(\mathbf{p}),$$

where $m_i(\mathbf{p})$ is a monomial of the coordinate components of $\mathbf{p}$ of degree k or less, and a_i are coefficients. For example, a constant, linear and quadratic polynomial in 2D has the form $g(\mathbf{p}) = a_1$, $g(\mathbf{p}) = a_1 + a_2 p_x + a_3 p_y$ and $g(\mathbf{p}) = a_1 + a_2 p_x + a_3 p_y + a_4 p_x^2 + a_5 p_y^2 + a_6 p_x p_y$, respectively, where p_x, p_y are the coordinates of $\mathbf{p}$. The monomial coefficients a_i are what we will need to find.

Note that now we have more unknowns (n bump coefficients c_i plus l monomial coefficients a_i) than the number of interpolation

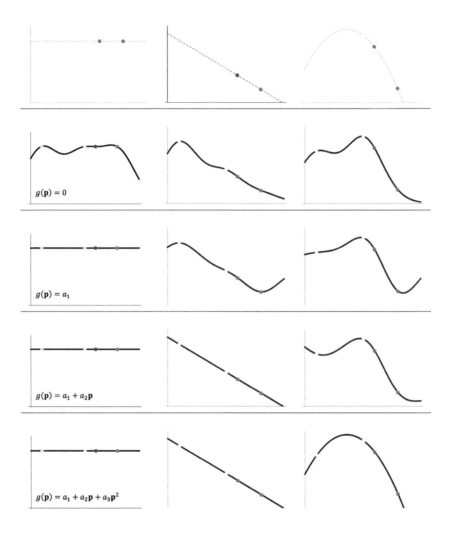

Figure 10.10: RBF interpolation using equation (10.8) for data sampled from a constant function (left), linear function (middle) and quadratic function (right). The different rows show results without using the polynomial term $g(\mathbf{p})$, and using $g(\mathbf{p})$ with 0,1 and 2 degree.

constraints (n). To constrain the system, we will add one more orthogonal constraint for each monomial term in $g(\mathbf{p})$,

$$\sum_{i=1}^{n} c_i \, m_j(\mathbf{p}_i) = 0\,, \quad \text{for } j = 1, \ldots, l\,. \tag{10.9}$$

These additional constraints will enforce the interpolating function to have square integrable second derivatives. Putting together the interpolation and orthogonal constraints, the linear system we will solve has the form

$$\begin{bmatrix} A & M \\ M^\top & 0 \end{bmatrix} \begin{bmatrix} \mathbf{c} \\ \mathbf{a} \end{bmatrix} = \begin{bmatrix} \mathbf{v} \\ 0 \end{bmatrix}\,, \tag{10.10}$$

where $A, \mathbf{c}, \mathbf{v}$ are defined in Equation (10.7) and

$$M = \begin{bmatrix} m_1(\mathbf{p}_1) & \cdots & m_l(\mathbf{p}_1) \\ \vdots & \ddots & \vdots \\ m_1(\mathbf{p}_n) & \cdots & m_l(\mathbf{p}_n) \end{bmatrix}\,,$$

$$\mathbf{a} = [a_1, \ldots, a_n]^\top.$$

The last three rows in Figure 10.10 compare the augmented RBF interpolation using polynomials of degree 0, 1 and 2. Observe that the use of a k-degree polynomial term $g(\mathbf{p})$ reproduces all functions up to degree k. For example, a linear polynomial (fourth row) is capable of reproducing the constant and linear function from which the data is sampled (first two columns), but not a quadratic function (last column).

Practical issues in surface reconstruction. The RBF technique, although discussed above only in 1D and 2D, can be performed in 3D and higher dimensions in the same fashion. Due to the smoothness of the resulting function, RBF has been one of the most popular choices for surface reconstruction from scattered points (see Figure 10.2(b)). However, as the size of input (e.g., the number of scattered points) grows, practical issues such as efficiency and stability arise. First, solving the linear system in Equation (10.10) and evaluating the interpolated value $f(\mathbf{p})$ using Equation (10.8) becomes time-consuming as the input exceeds thousands of points, which is quite common nowadays. Second, the linear system in Equation (10.10) becomes poorly conditioned as the size and irregularity in the input increases, meaning larger errors are likely to appear in the numerical solutions (see more discussions on linear solvers in Chapter 6).

There are two general approaches to make RBFs more practical, especially in surface reconstruction from large and noisy point sets. The method of fast RBF [CBC+01] utilizes mathematical tools such as matrix pre-conditioning [BCM99] and the fast multipole method (FMM) [BL97] to be able to efficiently solve and evaluate RBF. More commonly, radial functions with compact support are used to reduce the linear system in Equation (10.10) into a sparse system containing zero in most entries, which is more efficient and stable to solve. Unlike those shown in Figure 10.8, a compactly supported radial function $\phi(r)$ evaluates to zero except for a very small region around $r = 0$. The use of compactly supported functions often needs to be coupled with hierarchical evaluation approaches [OBS03] with varying support size in order to "fill holes" in non-uniformly sampled data, as shown in Figure 10.11 (left). For details of these two approaches, we refer interested readers to the papers [CBC+01, OBS03].

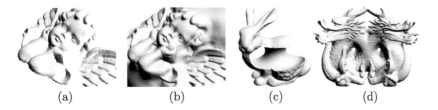

(a)   (b)   (c)   (d)

Figure 10.11: (a) A non-uniformly sampled point cloud and (b) its RBF reconstruction. CSG operations on the RBF reconstruction, including (c) subtracting a torus from the bunny and (d) adding two dragons. The figures are courtesy of Ohtake et al. [OBS03].

Another advantage of using an RBF for surface reconstruction is that it offers an implicit function $f(\mathbf{p})$ whose sign indicates whether $\mathbf{p}$ lies inside and outside of the surface. Certain mesh processing operations, such as CSG (constructive solid geometry) and offsetting, can be easily implemented given these implicit functions. Figure 10.11 (right) shows two typical CSG operations, adding and subtracting, on RBF-represented models. The resulting surfaces are represented by the zero-level set of a new implicit function whose values are the minimum or maximum of the input implicit functions at each evaluated point.

Chapter 11

Topology: How Are Objects Connected?

Niloy J. Mitra

Topology, in simple terms, gives us a tool to study and understand how an object is connected. Unlike differential geometric quantities, topology of an object changes only under extreme changes—when we allow tearing of the object, or punching holes into it. Hence, *topological invariants* are global quantities, and are more robust compared to local geometric measures.

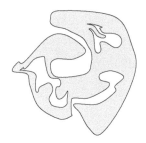 

Figure 11.1: Given 2D and 3D shapes, we are often interested in determining the number of holes present in such shapes. While for shapes like those on the left the answer may be clear, for complex shapes like the one on the right, the total number of holes may not be immediately obvious.

Let us start with an easy question: In Figure 11.1, how many holes are present in the 2D object on the left? How about the

Figure 11.2: Given curves or surfaces we are often interested to smoothly morph or deform one shape into another without tearing apart or gluing together parts of the shape. There are pairs of shapes for which such a deformation is not feasible. For example, while we can continuously deform the figure on the left to the one at the center, and vice versa; we cannot continuously deform the shape on the right to any of the other shapes. The intuitive explanation lies in the observation that the connectivity of the shape on the right differs from the other ones. As a result without changing the connectivity or performing topological surgeries once cannot morph the right figure to the other ones.

3D object on the right? Are the answers to such questions always obvious? To get some idea why this is a relevant question to ask, let us consider the following scenario: How can we smoothly deform one shape into another? Say for example, can we deform the shape on the left in Figure 11.2 to the shape in the middle, or to the one on the right? If we are talking about smooth deformations, where we cannot tear the object into parts or merge segments, then we cannot deform the one on the left to the one on the right. The simple explanation being that they are connected differently. So we cannot smoothly deform one to the other. The same argument applies to 3D objects (see Figure 11.3). Topology helps us to make such notions precise.

Figure 11.3: Can we smoothly deform the sphere on the left to the torus on the right? The answer is no. The objects are topologically different, i.e., they are connected differently and hence one cannot be morphed into the other without changing their connectivity.

Topology abstractly represents an object retaining only the essential components that deal with how an object is connected. To illustrate, we consider a classic problem: The Seven Bridges of Königsberg, where we find one of the early applications of topological concepts (see Figure 11.4). The city of Königsberg has four land masses $\{A, B, C, D\}$ inter-connected by seven bridges. The problem is to find a path, if any exists, that crosses each bridge *exactly* once. Using topological abstraction, Euler mapped the problem to a graph, and arguing on the degrees of the graph nodes, showed that such a path *cannot exist*. While we leave it to the reader to work out the details of the argument, this example demonstrates how useful connectivity can be. Let us now see why topology is an important computer graphics tool.

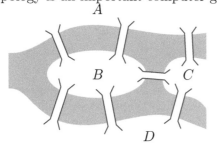

Figure 11.4: In the town of Königsberg the seven bridges connecting four landmasses A, B, C and D (left) is the center of an interesting question: Is it possible to walk between the four parts, crossing each bridge once and only once? Euler (right) proved that such a path cannot be extracted using an argument based on the connectivity defined by the arrangement of landmasses and bridges.

Suppose we are given a triangle mesh. Can it have an arbitrary number of vertices, edges, and faces? Is there any relation between them and the number of holes the object has? The answer is yes, there indeed exists such a special relation—the Euler equation, which we will learn about soon. Thus, we see how tightly coupled topological concepts are to geometric models.

In graphics, in animation, and in geometric modeling we are often interested in morphing or (smoothly) deforming a shape into another, like transforming a cat into a lion. However, if we want to go from a doughnut to a ball, topology tells us this is not possible unless we change the connectivity by tearing objects (Figure 11.3).

We will see that topological invariants are very strong, and not easy to change! On the other hand, in the presence of noise and outliers, points or surface parts can get disconnected, resulting in significant changes in surface connectivity. Such spurious connectivity can severely affect standard geometry processing tasks like noise cleaning, parameterization, subdivision, simplification (see Figure 11.5), etc.

Figure 11.5: An input model (left) with small topological holes (not visible in the figure) can lead to a non-intuitive simplification result (center). A smarter simplification that ignores small topological holes and tunnels gives a much better simplification result (right) for the same final triangle count. See [WHDS04] for more details.

As a last example, let us consider Boolean operations between two objects. Given two objects, say we are interested in their union, or in other words, the part of space occupied by both of the objects. For this purpose, we have to be able to differentiate between the inside and outside of an object. Topology tells us there exist (closed) objects with no notion of inside or outside! It also tells us about consistent orientation of an object. Classic examples include a Möbius strip (Figure 11.6) and a Klein bottle (Figure 11.7).

Fundamental topological concepts and invariants

Topology being such an important concept requires a slow and gradual exposure to fully learn, understand, and assimilate its var-

Figure 11.6: Möbius strip, discovered by August Möbius, is a classic example of a non-orientable surface. It is a surface (of zero thickness) with one side as can be seen by following the blue arrows shown on the figure on the right. As a mockup, one can construct a Möbius strip by starting with a paper strip, giving it a half twist, and then gluing together the ends of the strips.

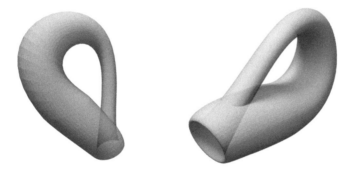

Figure 11.7: Klein bottle, originally introduced by Felix Klein, is another example of a non-orientable surface, i.e., a surface without a distinct notion of *inside* and *outside*. Although in 3D one cannot embed a Klein bottle without self-intersections, in a 4-dimensional space a Klein bottle can be embedded without self-intersection.

ious definitions and subtleties. Unfortunately, such an exposition is beyond the scope of this chapter. Instead, in the following we will just try to gain intuitive understanding of the very basic concepts, and ignore computational issues. The interested reader is encouraged to take her time to learn about these topics in greater depth.

Manifold. Let us get an informal understanding of manifolds, an important abstract topological concept. On a 1D manifold or one-manifold, at every point, the neighborhood has the same connectivity as that of a line. In other words, at every point, we have only two possible directions to go. For a 2D manifold, the neighborhood around each point looks like a disk or a plane. Lines, circles, or any simple curve are examples of 1D manifolds, while spheres and torii are examples of 2D manifold surfaces. If a manifold of dimension d can be embedded or placed in an Euclidean space $\mathbb{R}^n$, then $d \leq n$, i.e., the dimension of a manifold is always less than the dimension of its embedding space.

In geometric modeling, we often encounter non-manifold surfaces, especially when we have data corrupted with noise and outliers, say, as data obtained from a range scanner. In Figure 11.8 we see an example of a non-manifold surface. Around the point $\mathbf{p}_3$, the neighborhood has connectivity of a disk. However, along the edge $\mathbf{e}$ we have three triangle faces meeting. Thus, around points $\mathbf{p}_1$ and $\mathbf{p}_2$, topologically, the neighborhoods are *not* disks. So, we have a non-manifold surface.

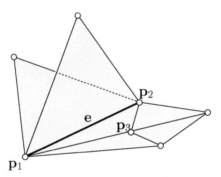

Figure 11.8: While the neighborhood around point $\mathbf{p}_3$ is topologically like a disc, the neighborhoods around points $\mathbf{p}_1$ and $\mathbf{p}_2$ are different. Edge $\mathbf{e}$ has three faces incident on it, and hence we have a non-manifold surface.

Boundary. Informally, a *boundary* is the edge of a surface. For a polygonal mesh, edges with only one incident face are boundary edges. The remaining edges, or the *internal* edges, have exactly two incident faces. Recall that if an edge is shared by more than two triangles, then we have a non-manifold surface (see Figure 11.8). In the example shown in Figure 11.9, the boundary edges are marked in yellow, while the internal edges are shown in gray. A manifold mesh without any boundary edges is referred to as a *closed manifold*, for example, the surface shown on the right in Figure 11.9.

Figure 11.9: A closed manifold mesh with the edges marked in gray (left). A manifold mesh with boundaries (right). The boundary edges are marked in yellow, while the interior edges are shown in gray. For a polyhedral model representing a manifold surface, any edge with only one incident face is a boundary edge.

Connectivity.   We are all familiar with the concept of *components*, or disconnected patches. Informally, a region of an object is said to form a component if it is possible to travel between any two points of the region along the surface of the object. Alternately, if we cannot travel from one point of the object to any other point of the same object by traveling along the object surface, we have an object with multiple components. For example, a sphere has one component, while a scene with two disjoint spheres has two parts or components. The number of components of an object, denoted by β_0, remains invariant as long as we are not allowed to tear apart the object, or glue together object parts. Higher-order numbers such as $\beta_1, \beta_2, \ldots$ capture other topological quantities to better characterize the connectivity of object models. These terms are called *Betti numbers*. The name was coined by Henri Poincaré in honor of Enrico Betti for his early contributions in topology.

Genus.   Next, let us talk about one of the earliest concepts in topology, the *genus* of an object. Informally, the genus is the

number of holes in an object. It is defined as the maximum number
of non-intersecting simple closed curves that can be simultaneously
drawn on an object such that the complement or the remainder
of the object remains singly connected. For example, any closed
curve drawn on a (hollow) sphere divides its complement into two
parts or components. Hence, the genus of a sphere is *zero*.

However, for a torus, it is possible to draw one closed curve
that still leaves the complement singly connected. Further, no two
disjoint closed curves can be simultaneously drawn on a torus such
that the complement still remains connected. Hence, the genus of
a torus is *one*, which also agrees with the informal understanding
that the torus has a single hole (see Figure 11.10). The genus,
often denoted by g, is also the number of handles on the object.

Figure 11.10: A torus is a closed manifold surface with genus one. It
is possible to cut the torus along a simple closed curve such that the
remaining surface is still singly connected, as shown in the right figure.
Further, it is not possible to have two disjoint closed curves such that
the complement remains singly connected. Hence, the genus of a torus
is one.

The number of *tunnels* or hollow cavities are of particular topological interest. They are denoted by β_2. For example, a hollow
or empty sphere has $\beta_2 = 1$, while a solid one has $\beta_2 = 0$. For a
hollow and solid torus, $\beta_2 = 1$ and $\beta_2 = 0$, respectively.

Orientability. A surface S embedded in a Euclidean space is
said to be *orientable* if a 2D figure starting from a point on S can
be continuously moved along the surface and back to the starting
position such that the 2D figure is in the same starting configuration. Otherwise a surface is said to be *non-orientable*. The Möbius
strip (Figure 11.6) and the Klein bottle (Figure 11.7) are classic
examples of non-orientable surfaces.

Table 11.1: The number of components β_0, the genus g, and the number of voids or tunnels β_2 for some common surfaces.

Surface	β_0	g	β_2
plane	1	0	0
hollow sphere	1	0	1
hollow torus	1	1	1
solid torus	1	1	0

In the discrete setting, say for a triangle mesh, the notion of orientability is even simpler. We can assign an orientation (clockwise or anti-clockwise) to any triangle of a given mesh, more generally to any face of a given polyhedron. Now two faces sharing an edge are said to be *consistently orientated* if the orientations of the two faces are opposite along the common edge. If an assignment of face orientations exists such that all neighboring faces of the given polyhedron are consistently orientated, then the given polyhedron is said to be orientable, otherwise it is non-orientable.

Euler characteristic. Finally, we learn about the *Euler characteristic*, χ, a commonly used topological invariant, i.e., an entity that remains constant as long as the connectivity or topology of the object remains unchanged. We note that the Euler characteristic is related to a few important topological invariants, most of which are beyond the scope of this book. Let us study the Euler characteristic in the context of polygonal meshes, more specifically triangle meshes. Suppose a triangle mesh P has V vertices, E edges, and F faces. Then the Euler characteristic is given by

$$\chi(P) = V - E + F.$$

An amazing relation is as follows: irrespective of the triangulation (or, in general, any polyhedron), an (orientable) closed surface with genus g always satisfies the invariant

$$\chi(P) = 2(1 - g).$$

For example, for any given triangulation of a sphere ($g = 0$), the numbers of vertices, edges, and faces always satisfy the special relation, $V - E + F = 2$. For a torus, which has genus one, we have $V - E + F = 0$. This may seem rather surprising, so let us try to study some specific examples and understand how this invariant is maintained.

Suppose we start with a simple tetrahedron. Topologically, a tetrahedron is equivalent to a sphere, i.e., one can smoothly deform a tetrahedron into a sphere, and vice versa without tearing or gluing it (Figure 11.11). Since the genus of an object is only related

Figure 11.11: A tetrahedron can be smoothly morphed or deformed into a sphere without performing any topological surgeries.

to the topology or connectivity of that object, a tetrahedron and a sphere both have genus zero. Thus a tetrahedron's characteristic number is zero since $2(1 - 0) = 2$.

Now, a tetrahedron has four vertices, six edges, and four faces. So its characteristic number is $\chi = 4 - 6 + 4 = 2$, and the invariant we just studied holds true. Let us now subdivide one of the tetrahedron faces into three (see Figure 11.12). We introduce one new vertex, three new edges, and two additional faces since three new faces are created while the old face is destroyed. Now the *change* in characteristic function will be $1 - 3 + 2 = 0$, i.e., the invariant continues to hold. By a similar counting argument, one can show that any edge collapse leaves the invariant intact. This agrees with our understanding that topology deals with connectivity of an object, and not with its specific geometric realizations. No matter what the geometric realization is, such topological invariants are always satisfied!

Let us explore an application of the Euler characteristic. For a *closed manifold triangle mesh*, each edge has exactly two incident triangle faces, while each triangle has three edges. Thus, $2E = 3F$. Hence,

$$\chi(P) = V - E + F = V - 3/2F + F = V - F/2 = 2(1 - g).$$

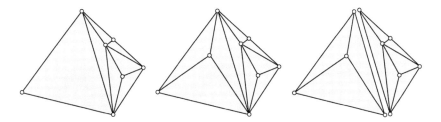

Figure 11.12: The Euler characteristic of a polygonal mesh P of genus g with V vertices, E edges, and F faces is defined as $\chi(P) = V - E + F$, and always equals $2(1-g)$. Comparing the left and the middle figures, we observe that for a triangle mesh, each step of a face-subdivision adds one new vertex, three new edges, and two additional faces, thus leaving $\chi(P)$ unchanged. By induction, we argue that any number of such subdivision steps keeps the Euler characteristic unchanged. It changes only due to topological surgeries that modify the genus of the surface, for example, due to a crack, as seen in the right figure.

For meshes with a large number of vertices and faces but a small genus, we can ignore g in comparison with V or F. So we have $F \approx 2V$, i.e., the number of faces is approximately twice the number of vertices for a triangle mesh representing a closed manifold surface. Similarly, we can argue $V \approx E/3$ or, $6V \approx 2E$ indicating that the average number of edges meeting at each vertex, i.e., the average valence of a triangle mesh, is six. Observe that such relations are meaningful only for large meshes, and not for ones like a tetrahedron mesh with only 4 vertices. We can use similar arguments for quad meshes where each face has four edges.

Topology in geometry processing

Surface reconstruction. Surface reconstruction deals with obtaining a complete watertight model from scanned data. Missing data and noise makes it a difficult problem, and at times even ambiguous or ill-posed. Recently, Ju et al. [JZH07] and Sharf et al. [SLS+07] (see Figure 11.13) demonstrated that a little user assistance to resolve topological ambiguities can be very effective to solve the problem. Intuitively, the user, by ruling out some topological possibilities or types of connectivity, gives valuable information about the family of shapes that could have generated the scanned data. Such hints or suggestions are sufficient to get a good watertight reconstruction from the scanned data.

Figure 11.13: Given an input model (left), a typical surface reconstruction method may have spurious regions, holes, or wrong connectivity (center). These errors mainly arise from topological ambiguity. Using user inputs for resolving such ambiguities, it is possible to get a much more desirable reconstruction (right). For details, please refer to [SLS+07].

Shape matching. In Chapter 9, we saw application of differential geometry to the shape matching problem. Topological invariants can also be very useful for solving the same problem. In shape morphing, as we saw earlier, if two objects are topologically different, then it is not possible to smoothly morph one into another since the objects have different global connectivity. Similarly, if two objects have different topology, then they are likely to be models of different objects, and hence unlikely to be a good fit when aligned or registered. So topological entities can be used as succinct descriptors to quickly rule out candidates for shape matching or object retrieval.

Reeb graphs are skeletal graphs capturing topological aspects of an object. Informally, a Reeb graph tracks how the connectivity of the cross-sectional curves of an object changes as we sweep an intersecting plane across the object. Such graphs can be efficiently constructed [PSBM07]. Further, Reeb graphs of objects are very easy to compare. Objects with different Reeb graphs or skeletal descriptors can be used for rejecting candidates for shape matching; this has been demonstrated to be an effective shape retrieval strategy by Biasotti et al. [BGSF08].

Concluding remarks

In this chapter, we got a glimpse of a few basic topological concepts, and applications of some topological concepts to various geometric modeling and processing tasks. The intention was to make the reader aware of the usefulness of such topological concepts and invariants, so that she may continue to study the concepts more thoroughly.

Before concluding this chapter, we would like to caution the reader about some robustness issues that we sometimes encounter while dealing with topological concepts. Topology only cares about global connectivity and questions like "Can we reach from one part of the object to another while staying on the surface?", "Are there holes in the object?", etc. Topology does not care about geometry, lengths of such paths, size of holes, etc. So two shapes that have vastly different geometry can still be topologically equivalent; and vice versa, two shapes that are almost identical in their entirety with some minor geometric differences can have vastly different topologies (see Figure 11.5). In spite of such geometric blindness, topological concepts can be very useful and powerful when used appropriately, especially in conjunction with suitable geometric measures.

Chapter 12

Graphs and Images

Ariel Shamir

Graphs play an important role in many computer science fields and are also extensively used in imaging and graphics. Often a major step in finding a solution to a problem in computer science amounts to formulating the problem using graphs. Informally, a graph is a set of objects called *nodes*, or *vertices*, connected by links called *edges* (Figure 12.1). Since graphs have been studied extensively, once a proper formalization for a specific problem is found, the algorithm to solve it can already be at hand. In this chapter, we will show how well-known graph algorithms can assist in segmentation of images, partitioning of 3D objects, and also in intelligently changing the size and aspect ratio of images and video while preserving their content [RSA08, SA09].

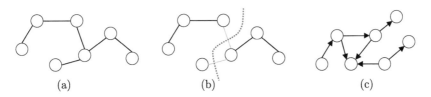

Figure 12.1: (a) A simple graph $G = \{V, E\}$ contains a set of nodes V (circles) and a set of edges E linking pairs of nodes. (b) A *cut* partitions V into two disjoint sets by removing a subset of edges. (c) In a directed graph, the edges are ordered pairs depicted by their direction.

Segmentation and region growing

Let us start with one of the fundamental problems in vision and image processing, namely, image segmentation. Segmentation strives to partition an image into sub-parts in some meaningful manner. It is considered one of the basic tasks of the human vision system which enables higher-level understanding of the world, and therefore is crucial for algorithms that seek semantic understanding of scenes. A segmentation procedure tries to give each pixel a value, also called a *label*, in order to group them in a meaningful manner. In Figure 12.2 one can see an example of partitioning a given image into regions with different labels using an algorithm called *watershed*. The algorithm uses a region-growing technique that operates on an undirected graph representation of the image. Region-growing algorithms are very useful in many applications. To understand region growing, we will first explain how an image can be represented as a graph.

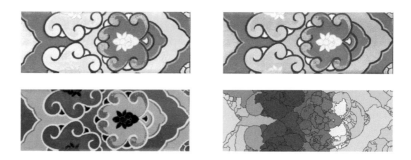

Figure 12.2: An example of segmenting an image into regions based on the watershed algorithm. The image is first converted into grayscale (top), and then inverted (bottom left) so that the boundaries between regions (i.e., the ridges between watersheds) will have the highest value (white). To prevent over-segmentation, all values below a certain threshold (gray level of 20) are considered minima of the function. Results are shown in the bottom right where each region is colored differently.

Region-growing. Formally, a graph, G, is defined as an ordered pair, $G = (V, E)$, where V is a (usually finite) set of *nodes*,[1] and E is a set consisting of pairs of elements from V, which are re-

---

[1]Nodes can be referred to as *vertices* as well, but we use the more general term *node* to distinguish them from vertices of meshes.

ferred to as *edges*.[2] If the edge pairs are ordered, then the graph is considered directed, otherwise it is undirected. A common way to visualize a graph is to draw the nodes as circles and the edges as segments connecting the two nodes defining the edge (see Figure 12.1). An image can be represented by a graph by defining each pixel as a node and linking it to its neighboring pixels with an edge (Figure 12.3).

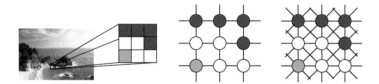

Figure 12.3: An image as a graph: each pixel is a node and neighboring pixels are connected with an edge. There are two types of pixel adjacency relations, which are named after the number of neighbors each pixel has: 4-connected adjacency (middle) and 8-connected adjacency (right). Waterfall image courtesy of Eric Chan.

To illustrate the use of this type of representation for images, we can start with a very simple algorithm called *flood-fill*. Assume you have a bounded region in an image. A *boundary* is defined as a connected subset of pixels sharing some common characteristic such as color (e.g., black pixels). A *bounded region* is a connected subset of pixels in the image that is completely enclosed by some boundary (or the image borders). The flood-fill algorithm colors a bounded region of a given image $\mathbf{I}$ in a given color. The algorithm starts at some specific pixel $\mathbf{I}(i, j)$ inside the region, where i and j are the indexes of the row and column in the image, and proceeds recursively as follows:

```
FloodFill( i, j ) {
   Pixel p = I(i, j)
   If not (stopTest(p) or isVisited(p)) {
      Visit(p)
      FloodFill(i, j + 1)
      FloodFill(i, j − 1)
      FloodFill(i + 1, j)
      FloodFill(i − 1, j)
   }
}
```

---

[2]To distinguish graph edges from visual edges found in an image, we will refer to the latter as *image edges*.

We assume that there is a storage space of one bit per pixel to mark if it was visited or not. The key component that defines the bounded region in the algorithm is the stopping criteria stopTest(p). This test will return true if p has the chosen boundary characteristic (e.g., its color is black). The name flood-fill comes from coloring the pixels in a predefined color inside the function Visit(p). This creates an effect of flooding the bounded region with this color (Figure 12.4). The four recursive calls in the algorithm above can be seen as recursive calls on all neighbors of a pixel using the 4-adjacency connectivity.

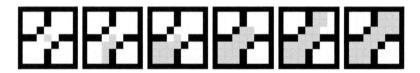

Figure 12.4: An example of filling a region using the flood-fill algorithm (image taken from Wikipedia Commons, created by André Karwath Aka).

Now that we view each pixel as a node in a graph where its neighboring pixels are connected to it by edges, we see that flood-fill is just a specific example of a more general algorithm on graphs that is called the *region-growing* algorithm. Using a general graph instead of the 4-adjacency graph created by images, we need to replace the four recursive calls with a loop on all the neighbors of a node. Thus, we obtain the general region-growing algorithm on graphs:

```
RegionGrow( Node v, Graph G ) {
    If not (stopTest(v,G) or isVisited(v)) {
        Visit(v)
        For all neighbors of u of v do
            RegionGrow(u, G)
    }
}
```

There are numerous examples for uses of the region-growing algorithms. In fact, simple depth-first search (DFS) is an example of the region-growing algorithm where we can visit each connected component of the graph as one region if we do not use any stopping criteria at all.

Watershed algorithm. Growing just one region is not enough to partition an image. Most images contain several regions and there is a need to enclose the basic region-growing algorithm in an additional loop to create multiple regions. Let us define a *free node* in a graph as a node that is not contained in any region. A full segmentation of an image can be created by representing the image as a graph and using the following graph-partitioning algorithm:

```
Partition( Graph G ) {
   Loop until there are no more free nodes in G {
      Choose a free node s as seed
      RegionGrow(s,G)
   }
   Clean up small regions
}
```

The resulting regions will be affected first by choosing different stopping criteria for growing a region. Instead of just looking at the pixel color, more complex characteristic functions can be used as stopping criteria. For instance, using a range or a set of color values, using the complement of a set of values, or even choosing some differential property of the pixels such as where the image gradient is large. The second factor that affects the partitioning results using this algorithm is how to choose the seeds for each region. Choosing different seeds can create different partitioning.

The *watershed* algorithm for image segmentation is just a partitioning algorithm where the seeds for regions and stopping criteria are chosen based on a definition of a height function. The watershed algorithm can be viewed as the process of flooding a landscape with water. Catchment basins are filled starting at each local minima. Wherever the water coming from different catchment basins meets, a watershed is found. These watershed lines separate individual catchment basins in the landscape (see Figure 12.5 for illustrations and Figure 12.2 for results of running the algorithm on an image).

Given a height function, the algorithm finds and labels all local minima of this function. Each minimum serves as the initial seed for a region. Next, all regions are grown incrementally from each

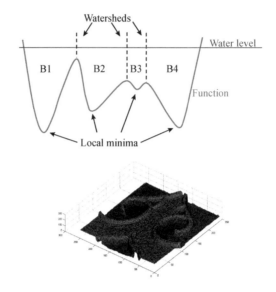

Figure 12.5: A 1D illustration of the watershed algorithm (top). The water level is gradually raised from all local minima and whenever two catchment basins meet, a watershed is found. A 2D illustration of part of the image from Figure 12.2 as a height function (bottom); one can see the catchment basins and watersheds.

seed until they reach a ridge or maxima in the function, thus partitioning the function terrain into regions (watersheds).

Using a priority queue based on the height function, this can be performed in one loop as follows:

```
Watershed( Graph G, function h ) {
   Find all local minima of h(G) as seeds {
   Create a priority queue Q
      the highest priority is the lowest height
   Label each seed with a unique label
   Insert all un-labeled neighbors of all seeds into Q
   Loop until Q is empty {
      Get the node v with highest priority from Q
      If all labeled neighbors of v have the same label
         Then label v with the same label
      Else
         Label v as a watershed node
      Insert all un-labeled neighbors of v to Q
   }
}
```

A key factor in using the watershed algorithm for image segmentation is the definition of the height function on the image. Several works defined various height functions based on the color or gray-level of the image, on differential properties such as gradients and more [VS91, BM93, MC08, CGNT09].

Graph region growing can be used in other domains as well. To illustrate this, consider a 3D object represented by a mesh. A mesh can also be viewed as a graph where the nodes are vertices and graph edges are the edges of each face (Figure 12.6(a)). However, one can also define a dual graph representation of a mesh, where the graph nodes are faces of the mesh, and two adjacent faces define an edge in the graph (Figure 12.6(b)). Similar to image segmentation, 3D object partitioning is also the subject of a major research effort, and region growing has also been used to partition a mesh or an object based on 3D stopping criteria [WL97, PKA03, ZH04, LDB05]. For example, Figure 12.6(c) shows a partitioning of a 3D mesh using region growing (image taken from [SSGH01]), where the dual graph of faces is used, and the stopping criteria is based on the normal difference between the faces in the growing region, along with a limit on the number of faces in each region.

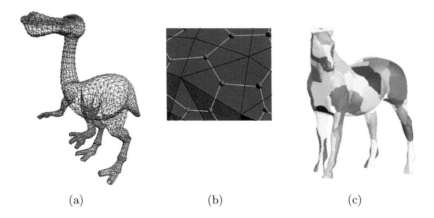

(a)   (b)   (c)

Figure 12.6: (a) A 3D boundary mesh is also a graph where vertices are nodes and the face edges are the graph edges. (b) One can also look at the dual graph of the mesh where each face is a node and there is an edge between adjacent faces. (c) Such dual graph is used for mesh segmentation based on a region-growing algorithm (image taken from [SSGH01] with permission).

Segmentation and graph cut

We have seen the advantage of using graph representations of images and meshes by utilizing graph region-growing to solve the segmentation problem. In this section a more direct translation will cast a segmentation problem into a graph partitioning problem.

It is often desirable to separate just a specific part of an image from the rest, for example, to separate a foreground object from the background image. This is used when cutting an object from one image and pasting it onto another. This problem can be viewed as labeling all pixels in the image either as foreground pixels (label 1) or background pixels (label 0). $L : \mathbf{I} \to \{0, 1\}$ will represent a binary labeling over all the pixels p of the image $\mathbf{I}$.

Segmentation as an optimization. The solution to this problem can be viewed as an optimization problem that minimizes some energy function. In our case, this energy depends on the labeling assignment of the pixels. First, the energy must measure the cost of assigning each pixel a given label. In addition, since the pixels of an object inside an image are usually connected, the energy also measures the cost of assigning neighboring pixels to different labels. For instance, if neighboring pixels have similar color but different labels, this should be penalized. Hence, the energy function mixes two components with some constant weighting factor λ:

$$E(L) = D(L) + \lambda S(L) \,. \qquad (12.1)$$

The first component $D(L)$ is called the *data* (or regional) term and describes the cost of assigning each pixel a foreground or a background label:

$$D(L) = \sum_{p \in \mathbf{I}} \text{cost}(p, L(p)) \,. \qquad (12.2)$$

$D(L)$ sums all the values associated with the assignment of a pixel p to the foreground ($L(p) = 1$) or to the background ($L(p) = 0$). This value is calculated based on the pixel's color value by comparing it to the foreground and background color models. These models usually contain more than a single color and are often based on Gaussian mixtures. The value of $\text{cost}(p, 1)$ is low when the color of a pixel is similar to the foreground color model, and cost$(p, 0)$ is low when the pixel's color is similar to the background model. These values will be high when the pixel's color is different from the corresponding model. Note that the greater the similarity of pixels to the correct color model, the lower the energy.

The second component $S(L)$ is called the *smoothness* (or boundary, contrast) term and describes the cost of assigning two neighboring pixels a different label:

$$S(L) = \sum_{p,q \in N} B(p,q) \cdot |L(p) - L(q)|, \qquad (12.3)$$

where N is the set of all neighbors in the image graph.

Note that $|L(p) - L(q)|$ can be either 0 if p and q have the same label, or 1 otherwise. The function $B(p,q)$ measures the color similarity between the two pixels. Its value is higher when p and q have similar colors. Hence, the greater the similarity of a pixel is to its neighboring pixels, the greater the energy term $B(p,q)$. However, this cost is added to the total energy only when the neighboring pixels are labeled differently. In other words, the smoothness term $S(L)$ sums up the cost of pixels similarity at the boundary between foreground and background. The main motivation of this term is to keep areas with similar colors in one segment and move the boundary between foreground and background to areas where color differences occur (i.e., image edges).

To calculate cost$(p, L(p))$ we need to define the similarity of a pixel's color to the foreground and background color models. This can be done in several ways. Very often in interactive applications, the user marks some pixels as foreground, and some other pixels as background (see Figure 12.7), and a color appearance model is created based on the two sets of marked pixels (for instance, using color histograms or Gaussian mixture models; see [MP00, RKB04]). The pixels that are chosen by the user to be foreground (or background) are added to the energy function as "hard" constraints, meaning they cannot change their labeling (for instance, if p is marked as foreground, then cost$(p, 1) = 0$ and cost$(p, 0) = \infty$).

The solution space for minimizing the energy function $E(L)$ is exponential, or 2^n if n is the number of pixels. Hence, a simple brute-force search for the minimum energy is not feasible. However, utilizing the graph representation of images yields a polynomial algorithm that can find the global optimum. The basic idea is first to build a specific weighted graph representation of

Figure 12.7: Marking the foreground with red and background with blue strokes (left), and segmenting part of an image (right) using graph cut.

the problem, and then use the well-known graph-cut algorithm for partitioning the graph.

Graph definition. In general, any partitioning of the nodes of a graph into two disjoint subsets (sub-graphs) is called a *cut*. On a connected graph, a cut can be achieved by removing a subset of edges from the graph. The set of removed edges comprises the cut (see Figure 12.1(b)). On weighted graphs (graphs whose edges carry weights or costs), each cut also carries a cost, which is the sum of the weights of the set of edges comprising the cut.

We use the 4-adjacency graph representation of an image where each pixel is represented by a node and each node is connected by an edge to its top, bottom, right and left neighbors. These edges are called *n-links*. We add two special nodes called *terminals*: the F-terminal represents the foreground and the B-terminal represents the background. We connect all pixel nodes to both terminals. These edges in this graph are called *t-links* (see Figure 12.8 (left)).

The main idea is to find a cut in the graph that separates the two terminal nodes. Such a cut defines two sub-graphs: one connected to the F-terminal and the other to the B-terminal (Figure 12.8 (right)). The nodes that are connected to the F-terminal are considered as foreground and the nodes that are connected to the B-terminal are considered as background. Using a simple mapping of the weights of the graph to the components of the

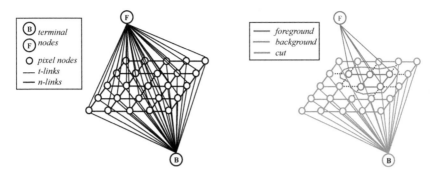

Figure 12.8: The graph created from an image for segmenting before (left) and after (right) the graph cut.

energy function $E(L)$ allows us to look for the *minimum cut* that separates the two terminal nodes. This cut also minimizes the energy function $E(L)$ and provides a global optimum for our original foreground–background separation problem.

The t-links' weights are inversely proportional to the data term. Every terminal represents the color model of the background or foreground. The more similar a pixel's color is to the terminal's color model, the greater the weight of the t-link connecting the node to this terminal will be, deferring the minimal cut from passing through this edge. The n-links' weights are proportional to the smoothness term. More precisely, if p and q are neighbors, we use the value $B(p, q)$ as the weight of the n-link between them. Neighboring pixels that have similar colors will have a larger weight on the edges connecting them, again, preventing the minimal cut from passing through these edges.

Note that a minimal cut in this graph will force all nodes to be connected to one, and only one, of the terminals. First, a cut forces every node to be connected to only one of the terminals, or else there is a path from one terminal to the other terminal through this node. Second, if a cut removes both t-links from a node, it will not be a minimal cut. This is because we can remove one of these t-links from the cut and add it back to the graph. The cut will still be valid, but its cost is smaller than the original cut. Hence, finding a minimum cut in the graph that we defined leads to the desired labeling: all nodes are either connected to the F-terminal and are labeled as foreground, or connected to the B-terminal and are labeled as background. By examining the energy terms it can also be shown that this labeling is optimal [GPS89].

Finding the minimum cut. We have converted the problem of minimizing an energy function that depends on the labeling of pixels to the cutting of a graph. However, there is an order of $2^{|E|}$ possible cuts, where $|E|$ is the number of edges in this graph and therefore an exhaustive search is still not feasible. However, the minimum cut problem in graphs is a well-studied problem, and there are several algorithms that solve this problem in polynomial time complexity.

A classic approach comes from *max-flow min-cut theorem* [FF62]. We look at the weights of the edges as capacities. A *capacity* is the maximum amount of flow that can pass through an edge from one side to the other (according to its direction). A flow is *valid* in a graph if no edge contains a flow above its capacity, and the amount of flow entering each node through its incoming edges is smaller than or equal to the amount of flow leaving it through its outgoing edges. Given a directed graph with two nodes denoted as source and sink, the maximum flow problem is defined as finding the maximum valid flow that can pass through the graph leaving the source and arriving at the sink.

The max-flow min-cut theorem states that the maximum flow from the source to the sink equals the cost of the minimum cut that separates these two nodes in the same graph. Moreover, all *saturated edges*, meaning all edges whose capacity is equal to their flow, compose the minimum cut. In our case, we use the F-terminal and the B-terminal as source and sink. We further direct all t-link edges from source to sink and replace all n-links with two directed edges with equal capacity as the n-link weight. Then, by solving the maximum flow problem we arrive at the minimum cut, which in turn will impose an optimal segmentation.

One of the most popular algorithms to solve the maximum flow problem is the Ford–Fulkerson algorithm [FF62]. This algorithm runs in $O(|E| \max f)$ where f is the amount of flow. The first step of the algorithm is to find an augmenting path, which classically is the shortest path between the source and the sink, with nonzero capacity edges on it (see Figure 12.9). The second step is to flood this path with an amount of flow that is equal to the lowest edge capacity among all the edges in the path. This process is called *augmentation*. The next step is to calculate the new capacity for each of the edges that participated in the path. Their new capacity will be their old capacity minus the amount of flow passing through

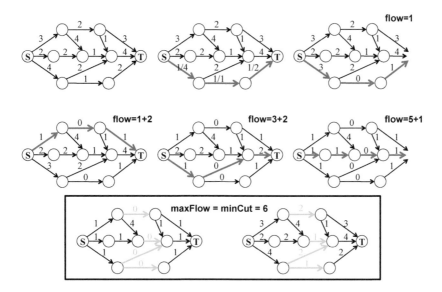

Figure 12.9: The Ford–Fulkerson algorithm. The top two rows illustrate finding augmenting paths in the graph until a maximum flow is reached. At the bottom, all saturated edges compose the cut, and the sum of their original weights is the maximum flow and the minimum cut.

them. This new capacity is called the *residual capacity*, and the new graph after calculating all the residual capacities is called the *residual graph*. Note that augmentation will cause at least one of the edges to be saturated, meaning its residual capacity will be zero. After the calculation of the residual graph, the algorithm returns to the first step to search for an augmenting path on the residual graph, flood it again and calculate a new residual graph. These iterations continue until no augmenting path can be found:

```
Ford--Fulkerson( graph G,node s, node t ) {
   Initialize f(e) = 0 for all e ∈ E.
   Repeat {
      Find an augmenting path A in G between s and t
      Augment f(e) for each edges e ∈ A
      G = calculate residual graph using A
   } Until no more augmenting paths found in G
}
```

The search for a new augmenting path can take $O(|E|)$ while in every iteration the capacity must be increased by one. This leads to the complexity of $O(|E| \max f)$. Variants of the classic algorithm include the Edmonds–Karp algorithm, which uses breadth-first search (BFS) to find the augmenting path with running time complexity of $O(|V| \cdot |E|^2)$, while the algorithm of Dinitz and Goldberg–Tarjan (Push-Relabel) solves the problem in $O(|V|^2 \cdot |E|)$, [CSRL01]. When dealing with mega-pixel images, the search for the shortest path in every iteration becomes prohibitive. It is tempting to replace the search for augmenting path with a simple greedy search for any unsaturated path (any path that can still carry some flow). However, such a scheme will not work in general, as illustrated in Figure 12.10.

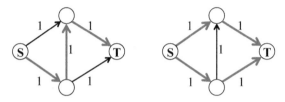

Figure 12.10: Greedily pushing flow into any unsaturated path in a graph may lead to sub-optimal flow. Choosing an arbitrary path from the source to the target results in a flow of 1, which is sub-optimal but cannot be extended (left). The optimal flow of 2 is shown on the diagram on the right. One method to find correct augmenting paths is to find the shortest path in the residual graph as in Figure 12.9.

Boykov and Kolmogorov [BK04] proposed a new algorithm to solve the maximum flow problem without the need to search for the shortest path. Their algorithm has a theoretical complexity which is worse than the above approaches, but in practice, on real images, this algorithm outperforms these approaches. The algorithm is based on two search trees, the S-tree and T-tree, both of which grow simultaneously from the source and the sink sides, respectively. Whenever the leaves of these two trees meet, a new path is found from the source to the sink. This path is then used for augmenting and a residual graph is calculated. It is important to note that the augmenting path found need not be the shortest one. Any path found will lead to an augmentation and consequently to an improvement of the amount of flow in the graph. This algorithm requires more iterations relative to algorithms that find the shortest path, but each iteration is much faster as any new augmenting path found is used to improve the flow.

The algorithm defines three kinds of nodes. The first are *free nodes*, which do not belong to any of the two search trees. The second are *active nodes*, which belong to one of the trees, have neighbors which are free nodes, and the edge connecting them is non-saturated. Free nodes can be acquired by the active nodes as new children in one of the trees. When a free node is acquired, it is added as the child of the active node that acquired it. The third are the *passive nodes*, which are blocked by nodes from the same tree and cannot become involved in the growing process of the tree. Using these definitions, the algorithm proceeds as follows:

```
Boykov-Kolmogorov( graph G,node s, node t ) {
    Initialize S = {s}, T = {t}, A = {s,t}, O = ∅
    Repeat {
        Grow S or T to find an augmenting path from s to t
        If an augmenting path P is found {
            AugmentOn(P)
            AdoptOrphans(T,S)
        }
    } Until no more augmenting paths found
}
```

The algorithm contains three main stages. The first stage is the *growing stage* in which the active nodes of each tree explore non-saturated edges connecting to free nodes. The newly acquired nodes become active nodes of the tree. Any active node that has no new node to acquire becomes passive. This stage terminates when an active node connects to a node from the other tree. This means a new augmenting path between the source and the sink is found (Figure 12.11). The second stage is the *augmentation stage* in which the path found in the growing stage is augmented. The minimal capacity of the path is subtracted from all of the edges of the path. This leads to a saturation of at least one edge. Consequently, at least one *orphan node* is created—a node that is not connected to any tree because the edge connecting it is saturated. More than one orphan node can be created and a forest can also evolve because of the augmentation process.

The goal of the last stage, the *adoption stage*, is to reconstruct two single trees emanating from the source and the sink, respectively. After this stage is done, the algorithm iterates back to the

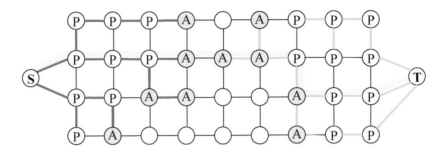

Figure 12.11: Illustration of Boykov–Kolmogorov max-flow algorithm. The S-tree (red) and T-tree (green) built on a graph representing an image. The active nodes are marked (A), passive nodes (P), and free nodes are empty. A possible augmenting path is marked in yellow—note that only one edge connecting the two trees creates a full path.

first stage. In the adoption stage, all orphan nodes try to find a valid parent from their original tree. A valid parent is an active node that has an edge connecting to the orphan node that is non-saturated. If such a parent is found, the node is added back to the tree by adding this edge. If no valid parent is found for an orphan node, then the node becomes a free node and can be acquired by both trees in the next stages. All the neighboring nodes of a newly created free node, that also have a non-saturated edge connecting it to them, become active nodes again. Note also that during this process, nodes can switch from one tree to another. The efficiency of this algorithm, compared to previous approaches, comes from the fact that the search trees are reused between iterations, and there is no need to reconstruct a whole new path in each iteration.

Extensions for multi-way graph cut and multi-label segmentations also exist in [BV06]. The graph-cut algorithm is also used for mesh segmentation, see, e.g., [KT03, LSTS04, JLCW06, SSCO08].

Image retargeting and dynamic programming

In the previous section we saw how the careful definition of a graph reduces an optimization problem to a known graph algorithm. In this section we will see more examples, first for solving a complex 2D problem using an efficient algorithm, and then extending the 2D solution to 3D. The problem we deal with is changing the size of media non-uniformly (i.e., changing its aspect ratio), which is often called *retargeting* (see [SA09]). Images and videos are displayed today on various devices. To accommodate high-definition TVs, computer screens, PDAs and cell phones, not only the resolution, but often the aspect ratio of the media must be adjusted. Standard image scaling techniques are not sufficient since they are oblivious to the image content, and can typically be applied only uniformly, preserving the image aspect ratio; otherwise stretching or shrinking may appear. Cropping of the media is limited since it is constrained to removing pixels from the image fringes, and it cannot support the expansion of the image. More effective resizing solutions can be achieved by considering the image content and not only shape constraints (Figure 12.12).

Figure 12.12: Non-uniform resizing changes the aspect ratio of the image: the top image is resized to half of its original width (bottom images). Compare (from the left) the use of *cropping*, which loses information, *scaling*, which creates shrinking artifacts, to content-aware resizing using *seam carving*.

Formally, the problem of image retargeting can be stated as follows. Given an image $\mathbf{I}$ of size $(n \times m)$, we would like to produce a new image $\mathbf{I}_r$ of a new size $(n_r \times m_r)$ that will be a faithful representation (as much as possible) of the original image $\mathbf{I}$, although either $n_r \neq n$ or $m_r \neq m$ (or both). There is no clear definition or measure of the fidelity of $\mathbf{I}_r$. In loose terms we would like the *content* of $\mathbf{I}$ to be preserved and $\mathbf{I}_r$ not to contain any visual artifacts.

Many *content-aware* retargeting techniques have been presented in literature and most follow a common two-stage approach to the problem [SS09]. First, an *importance* or saliency map is defined on the original image **I**, protecting its important content. Second, some resizing operator is applied to **I** that considers the importance map defined on **I**. Such methods enable, as far as possible, the adjusting of images and video to different aspect ratios with less distortion of the more important parts, or content, of the media. In this section, we present one of the discrete techniques for image retargeting called *seam carving* and show how it applies to images. Then, we will show how this operator can be realized using graph cut and utilized to retarget video as well.

The seam-carving operator [AS07]. The basic approach of seam carving for retargeting is to remove some pixels from the image in a judicious manner. We will assume that some saliency map is provided or calculated, and concentrate on the operator itself. For instance, since the human-vision system is more sensitive to edges, one can give a high importance value to image edges and low importance value to smooth areas using the following simple L_1 gradient-magnitude energy function on the image pixel $\mathbf{I}(i, j)$, which we will call E_1 (see Figure 12.13 (left)):

$$E_1 = e(\mathbf{I}(i, j)) = \left| \frac{\partial}{\partial x} \mathbf{I}(i, j) \right| + \left| \frac{\partial}{\partial y} \mathbf{I}(i, j) \right|. \qquad (12.4)$$

A *vertical seam* is a connected path of pixels in the image from the top to the bottom, containing one, and only one, pixel in each row of the image (see Figure 12.13 (right)). Formally, we define a

Figure 12.13: Color mapping of the L_1 gradient-magnitude energy used as a saliency map (left). The vertical cumulative cost matrix using Equation (12.8) (middle). A vertical and horizontal seams found on the image (right).

vertical seam as

$$\mathbf{s}^x = \{s_i^x\}_{i=1}^n = \{(i, x(i))\}_{i=1}^n, \quad \text{s.t. } \forall i, \; |x(i) - x(i-1)| \leq 1.$$
(12.5)

Note that x is a function $x : [1, \ldots, n] \to [1, \ldots, m]$ that is continuous in a discrete sense, where two consecutive values of the function do not differ by more than 1. Similarly, if y is a discretely continuous function $y : [1, \ldots, m] \to [1, \ldots, n]$, then an image *horizontal seam* is

$$\mathbf{s}^y = \{s_j^y\}_{j=1}^m = \{(y(j), j)\}_{j=1}^m, \quad \text{s.t. } \forall j, \; |y(j) - y(j-1)| \leq 1.$$
(12.6)

The pixels of the path of seam $\mathbf{s}$ (e.g., a vertical seam $\{s_i\}$) will therefore be $\mathbf{I_s} = \{\mathbf{I}(s_i)\}_{i=1}^n = \{\mathbf{I}(x(i), i)\}_{i=1}^n$. If we *carve out* or remove the pixels of one seam from an image, the effect is similar to the removal of a single row or column of pixels. First, the image height (using horizontal seams) or width (using vertical seams) is reduced by one. Second, this removal has only a local effect. In essence, all the pixels of the image are shifted up (or left) to compensate for the missing seam path pixels. This means that the visual impact is noticeable only along the path of the seam, leaving the rest of the image intact.

The key question in the seam-carving algorithm is how to choose the right seams to remove. We define the cost of a seam as the sum of energy of its pixels $E(\mathbf{s}) = E(\mathbf{I_s}) = \sum_{i=1}^n e(\mathbf{I}(s_i))$. Then, we can look for the optimal seam $\mathbf{s}^*$ that minimizes this cost:

$$\mathbf{s}^* = \arg\min_{\mathbf{s}} E(\mathbf{s}) = \arg\min_{\mathbf{s}} \sum_{i=1}^n e(\mathbf{I}(s_i)).$$
(12.7)

It is easy to see that there are an exponential number of possible seams, and therefore exhaustive search is impractical. However, it turns out that the characteristics of this optimization problem enable the use of a very efficient algorithm to find the optimal seam called *dynamic programming*. To use dynamic programming, some conditions must be met. First, the solution of the optimization problem must be obtained by some combination of optimal solutions of sub-problems. Second, the space of these sub-problems must be small, meaning that the solution of the original problem can reuse solutions of similar sub-problems many times.

In our case, if we can find, for every pixel in row $i - 1$, the optimal vertical sub-seam from the first row to that pixel, then we can find the optimal vertical seam from the first row to any pixel in row i by extending each sub-seam by one additional pixel. For each pixel in row i, e.g., $\mathbf{I}(i,j)$, the optimal seam can be extended from only three pixels because of connectivity: it can be extended from pixel $\mathbf{I}(i-1,j-1)$, pixel $\mathbf{I}(i-1,j)$, or pixel $\mathbf{I}(i-1,j+1)$ (see Figure 12.14 (top)). Since we know the optimal seams to all these pixels, we can choose the one with the smallest cost. Consequently, the cost of the optimal seam through pixel $\mathbf{I}(i,j)$ will be calculated as

$$
\begin{aligned}
\text{cost}(i,j) = e(i,j)+ \\
\min\big(\text{cost}(i-1,j-1), \text{cost}(i-1,j), \text{cost}(i-1,j+1)\big).
\end{aligned} \quad (12.8)
$$

Generalizing this observation, the optimal vertical seam is found by traversing the image from the first row to the last row composing a cost matrix C of size $(n \times m)$. For the first row we define $C(1,j) = e(1,j)$, and for each $i > 1$ we use Equation (12.8) to fill the matrix. At the end of this process, $C(i,j)$ will contain the cumulative minimum energy for all possible connected seams starting at the first row and ending at pixel $\mathbf{I}(i,j)$, and the minimum value of the last row in C will indicate the end of the minimal connected vertical seam (see Figure 12.13 (middle)). Hence, to find the path of the optimal seam we start from the minimum entry of the last row. Then we backtrack and follow, for each entry, the minimum entry of its three neighbors in the row above. For horizontal seams, the matrix C is composed similarly but from the first column to the last.

To reduce the width of an image by n pixels, we need to apply the seam-carving operator n times. That is, in each step we find an optimal seam in the image, remove the pixels associated with it and repeat the process n times. In fact, this approach resembles the dynamic shortest paths problem [RZ04], where finding the seam is equivalent to finding a shortest path on a graph, and removing the seam modifies the graph for the next iteration of shortest-path search.

Forward-looking energy [RSA08]. Choosing to remove the seam with the least amount of energy from the image (Equation (12.8)), works in many cases, but ignores energy that is *inserted* into the resized image *after* the seams are removed. This inserted

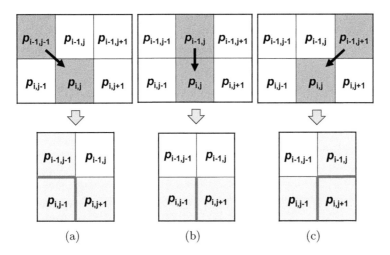

Figure 12.14: The three possible vertical seam steps for pixel $p_{i,j} = \mathbf{I}(i,j)$ (top). Calculating the step costs using *forward* energy (bottom). After removing the seam, new neighbors (in gray) and new pixel edges (in red) are created. In each case the cost is defined by the forward difference in the newly created pixel edges. Note that the edge between the new neighbors in row $i - 1$ is accounted for in the cost of the previous row pixel and therefore is not part of the cost of the seam through $p_{i,j}$.

energy is due to new intensity edges created by previously non-adjacent pixels that become neighbors once a seam is removed. Following this observation, a new *forward-looking* criterion is formulated. At each step, a seam whose removal inserts the minimal amount of energy into the image is sought. Such seams are not necessarily minimal in their energy cost, but leave fewer artifacts in the target image, after removal.

As the removal of a connected seam affects the image and its energy, only in a local neighborhood, it suffices to examine a small local region near the removed pixel. The energy introduced by removing a certain pixel is considered to be the cost of new *intensity edges* created in the image. This cost is measured as the differences between the values of the pixels that become new neighbors, after the seam is removed. Depending on the connectivity of the seam, three such cases are possible (see Figure 12.14). For each of the three possible cases, the cost is respectively defined as

(a) $\quad C_{\text{left}}(i,j) = |\mathbf{I}(i,j{+}1) - \mathbf{I}(i,j-1)| + |\mathbf{I}(i{-}1,j) - \mathbf{I}(i,j{-}1)|\,,$

(b) $\quad C_{\text{up}}(i,j) = |\mathbf{I}(i,j+1) - \mathbf{I}(i,j-1)|\,,$

(c) $\quad C_{\text{right}}(i,j) = |\mathbf{I}(i,j{+}1) - \mathbf{I}(i,j-1)| + |\mathbf{I}(i{-}1,j) - \mathbf{I}(i,j{+}1)|\,.$

The forward-looking cumulative cost matrix CF is used to calculate the seams using dynamic programming very similar to the matrix C from Equation (12.8). For vertical seams, each entry $CF(i, j)$ is filled from top left to bottom right using the following rule:

$$CF(i,j) = P(i,j) + \min \begin{cases} CF(i-1,j-1) + C_{\text{left}}(i,j), \\ CF(i-1,j) + C_{\text{up}}(i,j), \\ CF(i-1,j+1) + C_{\text{right}}(i,j), \end{cases} \quad (12.9)$$

where $P(i, j)$ is an additional pixel-based energy measure that can be used in addition to the forward energy cost. This energy can be the result of high-level tasks such as a face detector, or a user-supplied weight. Figure 12.15 shows a comparison between the results of changing an image size using the two formulations: backward and forward.

Figure 12.15: A comparison of results for reduction and expansion of the car image (top left) using the least-cost seam of Equation (12.8) (left in each pair) and the forward-looking (least inserted cost) seams of Equation (12.9) (right in each pair).

Image enlarging by seam insertion. The process of removing vertical and horizontal seams can be seen as a time evolution process. Denote $\mathbf{I}^{(t)}$ as the contracted image created after t seams have been removed from $\mathbf{I}$. To return to the original image, these

t seams must be inserted back into the image. Therefore, to enlarge an image, an approximate *inversion* of this time evolution process is used and new *artificial* seams are inserted into the image. Specifically, to enlarge the size of an image $\mathbf{I}$ by one, the optimal vertical (horizontal) seam $\mathbf{s}$ on $\mathbf{I}$ is computed, and simply duplicated. More careful insertion can be achieved by averaging the newly inserted pixels of $\mathbf{s}$ with their left and right neighbors (top and bottom in the horizontal case).

Denote the resulting, larger, image as $\mathbf{I}^{(-1)}$ using the time evolution notation. Unfortunately, repeating this process will most likely create a stretching artifact by choosing the same seam again and again (Figure 12.16(b)). To achieve effective enlarging, it is important to balance between the original image content and the artificially inserted parts. Therefore, to enlarge an image by k, the first k seams for *removal* are found, and duplicated to arrive at $\mathbf{I}^{(-k)}$ (Figure 12.16(c)). This can be viewed as the process of traversing back in time to recover pixels from a larger image that would have been removed by seam removals (although it is *not* guaranteed to be the case).

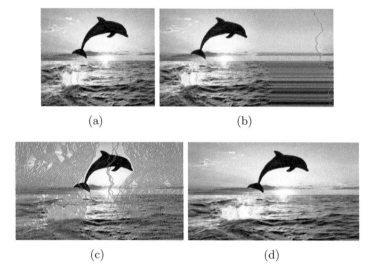

Figure 12.16: Seam insertion: Finding and inserting the optimum seam on an enlarged image (a) will most likely (b) insert the same seam again and again. (c) Inserting the seams in order of removal achieves (d) the desired enlargement.

Retargeting and graph cut

Similar to images, resizing videos for various devices is a challenging problem. However, the dynamic programming formulation that worked well for images in 2D does not scale well to 3D. Interestingly, the solution to this problem lays yet again in using a graph-cut formulation. For videos, it is often convenient to consider the sequence of frames as a 3D space-time cube [SSSE00, KSE$^+$03], and use voxels connected temporally (through time) instead of just pixels. The graph constructed is different than the one used for segmentation purposes. The main challenge is to design a 3D graph that produces a special type of cut, whose intersection with each 2D video frame (or image) produces a valid seam. To achieve this, the cut must satisfy two constraints:

Monotonicity: The seam produced must include one, and only one pixel, in each row (or column for horizontal seams).

Connectivity: The pixels of the seam must be connected.

The monotonicity constraint ensures the mapping is a function, while the connectivity constraint enforces continuity. Together, they ensure that the seam created from the cut would be valid and connected. Hence, the challenge is to construct a graph that guarantees the resulting cut will be a continuous function over the relevant domain. Standard graph cut-based constructions as seen in the previous sections do not satisfy these constraints as they can create arbitrary cuts.

For simplicity, let us first formulate the seam-carving operator for vertical seams in an image, as a minimum cost graph-cut problem. For horizontal seams, all constructions are similar with the appropriate rotation, while the extension to video will be presented later. In the new graph construction, every node still represents a pixel, and is connected to its neighboring pixels in a grid-like structure. For a 4-connected graph this means the neighbors are $Nbr(\mathbf{I}(i,j)) = \{\mathbf{I}(i-1,j), \mathbf{I}(i+1,j), \mathbf{I}(i,j-1), \mathbf{I}(i,j+1)\}$ (see Figure 12.3). However, for different formulations, different directed edges may be used and some diagonal edges connecting 8-connected neighbors may be used as well. The main difference from previous graph constructions is that the virtual terminal nodes, s (source) and t (sink), are connected *only* to the pixels of the leftmost and rightmost columns of the image, respectively, with infinite weight edges as shown in Figure 12.17. For video, these nodes are connected only to the sides of the video cube as shown in Figure 12.18.

A *cut* C in a graph partitions the nodes into two disjoint subsets S and T such that $s \in S$ and $t \in T$. All nodes to the *left* of the cut (connected to s) are labeled S and all nodes on the *right* of the cut (connected to t) are labeled T. In a directed graph, the cost of a cut only takes into account edges directed from nodes labeled S to nodes labeled T (these edges contain the flow). To define a seam from a cut, we consistently choose the pixels (nodes) immediately to the right of the cut edges (see Figure 12.17(b)). The optimal seam is then defined by the *minimum cut*, which is the cut that has the minimum cost among all valid cuts.

Following the L_1-norm gradient magnitude energy E_1 defined in Equation (12.4), the weights of the forward edges (from S to T) can be defined as the forward difference between the corresponding pixels in the image. In the horizontal direction, this will be $\partial x(i,j) = |\mathbf{I}(i, j + 1) - \mathbf{I}(i,j)|$, and in the vertical, $\partial y(i,j) = |\mathbf{I}(i + 1, j) - \mathbf{I}(i,j)|$. However, to ensure the cost of the cut is exactly the same as the cost we defined for seams in Equation (12.4), we assign the combined cost of the vertical and horizontal derivatives to the horizontal edges. The weight of the forward edges will be the combined weight of vertical and horizontal differencing or $E_1 = e(i,j) = \partial x(i,j) + \partial y(i,j)$ (Figure 12.17(a)).

To impose the monotonicity constraint on the cut, we add backward edges (from T to S) with infinite weight. The purpose of these infinity edges is to constrain the possible types of cuts because the optimal cut cannot pass through such edges. Backward infinity edges impose the monotonicity constraint as the optimal cut cannot cross an infinity cost edge from the source to the target, and therefore must cut each row only once (see Figure 12.17(c)). Similarly, to constrain cuts to be connected, infinite weight *diagonal* edges going *backward*, from T to S are used. This limits the cut from jumping more than one node (pixel) while moving to the next row (see Figure 12.17(d)).

A cut in such a graph is monotonic and connected. It consists of only horizontal forward edges (the rest are infinite weight edges that pose the constraints and cannot be a part of the cut). Hence, the cut cost is the sum of $E_1 = e(i,j)$ for all seam pixels, which is exactly the cost of the seam in the original seam-carving operator.

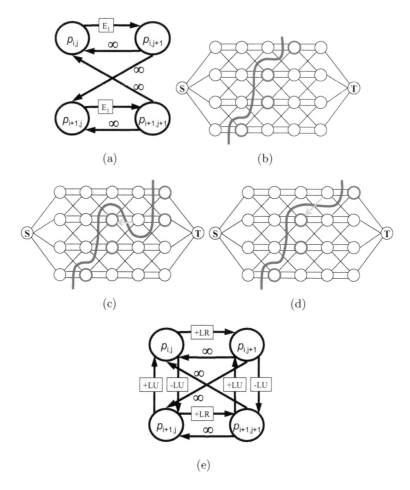

(a)

(b)

(c)

(d)

(e)

Figure 12.17: The graph construction for seam carving vertical seams using graph cut. The subgraph in (a) is tiled across the image while the left and right columns are connected with s and t, respectively, as illustrated in (b). Note that the backward horizontal infinity edges ensure that the cut is monotonic, or else an infinity edge from S to T must be cut (for example, the green edge in (c)). Backward diagonal edges ensure continuous seams, or else a diagonal infinity edge from S to T must be cut (for example, the green edge in (d)). Therefore, a cut in the full graph created from (a) is both monotonic and connected, and is equivalent to the optimal seam using Equation (12.8) (for example, using E_1 energy of Equation (12.4)). The sub-graph construction at (e) represents the forward energy of Equation (12.9) (see text for details).

Because both algorithms guarantee optimality, the seams created by both must have the same cost, and (assuming all seams in the image have different costs) the seams must be the same. Therefore, a cut in this graph defines a seam that is equivalent to the original dynamic programming algorithm. Note that energy functions other than E_1 that are defined on the pixels can be used as the weight of the forward horizontal edges. Using these energy costs as weights on the horizontal edge with graph cut guarantees the same results as seam carving with dynamic programming.

Defining the forward energy cost in a graph is somewhat more elaborate. There is a need to create a graph whose edge weights will define the cost of the pixel removal according to the three possible seam directions. Figure 12.17(e) illustrates this construction. A new horizontal pixel edge $p_{i,j-1}p_{i,j+1}$ is created in all three cases because $p_{i,j} = \mathbf{I}(i,j)$ is removed. Consequently, the difference between the **L**eft and **R**ight neighbors $+\mathbf{LR} = |\mathbf{I}(i,j-1)-\mathbf{I}(i,j+1)|$ is assigned to the graph edge between the nodes representing $p_{i,j}$ and $p_{i,j+1}$. To maintain the seam monotonicity constraint as before, $p_{i,j+1}$ and $p_{i,j}$ are connected with a (backward) infinite weight edge. Diagonal backward infinite edges are also added to preserve connectivity.

Next, we account for the energy inserted by the new vertical pixel edges. In the case of a vertical seam step (Figure 12.14(b)), there are no new vertical edges, so no energy is inserted. Since the cost of a cut will only contain edges directed from nodes labeled S to nodes labeled T, only upward vertical edges will point from S to T and will be part of the cut in a right-oriented cut (Figure 12.14(a)), and only downward vertical edges will point from S to T and will be part of the cut in a left-oriented cut (Figure 12.14(c)). Hence, the difference between the **L**eft and **U**p neighbors $+\mathbf{LU} = |\mathbf{I}(i-1,j) - \mathbf{I}(i,j+1)|$ is assigned to the upward vertical edge between $p_{i,j}$ and $p_{i-1,j}$, and the difference $-\mathbf{LU} = |\mathbf{I}(i-1,j)-\mathbf{I}(i,j-1)|$ is assigned to the downward vertical edge between $p_{i-1,j}$ and $p_{i,j}$ (note that $-\mathbf{LU}$ means the difference between the **L**eft and **U**p neighbors with respect to the endpoint of the arrow).

These graph constructions allow us to apply any seam-carving algorithm by performing a graph-cut operation. High-level functions, such as a face detector or a weight mask scribbled by the user, can also be used in any of these graph constructions. The

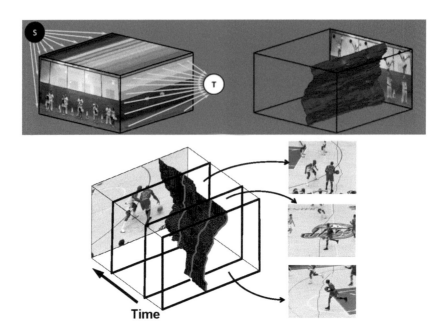

Figure 12.18: The graph construction on the video cube (top) guarantees that the graph cut will find a connected and monotonic manifold in the volume. The intersection of the manifold with each video frame defines the seam on the frame (bottom).

pixel's energy is simply added to the horizontal edge going out of the pixel.

The extension to video is straightforward and is described in [RSA08] (see Figure 12.18). Let us assume we are still searching for a vertical seam but on the video cube. In each frame, or $X \times Y$ plane, a graph that is similar to the graph used for a 2D image is used. Similarly, the same graph construction is used in any $X \times T$ plane in the cube including backward diagonal infinity edges for connectivity. The source and sink nodes are connected to all left and right (top/bottom in the horizontal case) columns of all frames, respectively. Now, a partitioning of the 3D video volume to source and sink using graph cut will define a manifold inside the 3D domain. Such a cut will also be monotonic in time because of the horizontal constraints in each frame that are already in place. This cut is globally optimal in the cube both in space and time. Moreover, restricted to each frame, the cut defines a 1D connected seam.

Chapter 13

Skewing Schemes

Daniel Cohen-Or

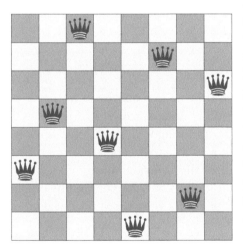

Figure 13.1: A possible solution to the eight queens problem.

In this chapter, we will show an example of the usefulness of number theory, at least for one of its known theorems. We start right away by discussing the well-known queens problem, whose solution does not require any use of number theory. The classic queens problem is to find the maximal number of queens that can be placed on a chessboard such that no two queens attack each other. We assume here that all the readers know how queens move and attack. The question then can be generalized to an $n \times n$ board, and then the challenge can be to find all possible solutions or, say, to generate one valid solution (see a possible solution in Figure 13.1). Such a problem can be easily solved by a simple recursive program, and indeed this problem is commonly given as an exercise in programming courses.

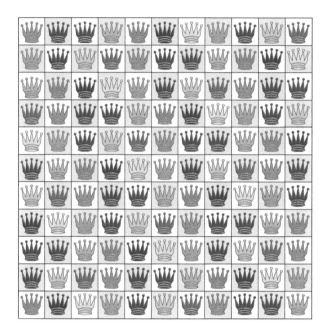

Figure 13.2: Eleven sets of eleven queens placed with no conflicts.

We are interested in a somewhat more challenging version of the problem. We are given an $n \times n$ chessboard over which we hope to place n sets of n queens such that no two queens from the same set attack each other. A solution is demonstrated in Figure 13.2. Note that in each row, column, and diagonal, each color appears only once. The ambitious reader can stop here and try to find more solutions, and to generalize the solution to any chessboard size. A generous hint was already given in the opening line of this chapter: number theory. This suggests that there is a scheme or an expression by which the n sets of n queens can be placed on an $n \times n$ board. Moreover, the problem and the solution can be easily generalized for a cubic board of $n \times n \times n$, over which we place n sets of $n \times n$ 3D queens that attack along all diagonals (including those that traverse between orthogonal slices) and rows across any axis.

Before moving on to show the solution to this problems, let us familiarized ourselves with the closely related notion of the *Latin square*: a Latin square consists of n sets of the numbers, 1 to n, arranged in such a way that no orthogonal (row or column) contains the same number twice. An 8×8 Latin square is demonstrated in Figure 13.3.

0	1	2	3	4	5	6	7
3	4	5	6	7	0	1	2
6	7	0	1	2	3	4	5
1	2	3	4	5	6	7	0
4	5	6	7	0	1	2	3
7	0	1	2	3	4	5	6
2	3	4	5	6	7	0	1
5	6	7	0	1	2	3	4

Figure 13.3: A Latin square, 8×8.

This definition can be extended to a *diagonal* Latin square, in which each of the symbols or numbers appears exactly once in each diagonal. The solution shown in Figure 13.2 is an example of a diagonal Latin square of dimension 11×11. Figure 13.4 displays such a diagonal Latin square, which is exactly the solution of the generalized queen problem that is displayed in Figure 13.2, where every queen color is translated into a unique index. Using this terminology, we are also interested in extending and solving a diagonal Latin cube.

Conflict-free access

The reader might rightfully wonder what the above has to do with practical applications in computer science. Well, these mappings of numbers into a square or a cube are in fact *skewing schemes*. To allow more efficient access to memory, especially to consecutive entries in two-dimensional or three-dimensional arrays (squares or cubes here) the memory can be partitioned into n memory banks or modules. This partitioning allows simultaneous, conflict-free access to n elements, one from each memory module. For example, it is possible to access in parallel the elements of an entire row or column of a matrix that is mapped to n memory modules by

0	1	2	3	4	5	6	7	8	9	A
7	8	9	A	0	1	2	3	4	5	6
3	4	5	6	7	8	9	A	0	1	2
A	0	1	2	3	4	5	6	7	8	9
6	7	8	9	A	0	1	2	3	4	5
2	3	4	5	6	7	8	9	A	0	1
9	A	0	1	2	3	4	5	6	7	8
5	6	7	8	9	A	0	1	2	3	4
1	2	3	4	5	6	7	8	9	A	0
8	9	A	0	1	2	3	4	5	6	7
4	5	6	7	8	9	A	0	1	2	3

Figure 13.4: The same scheme as in Figure 13.2 but with set indices rather than colored queens.

a Latin square scheme, as shown in Figure 13.3. Since diagonals are also an important access pattern in a matrix, we would like to develop a skewing scheme, like a diagonal Latin square, allowing conflict-free access to all diagonals. Some architectures for volume rendering were designed to allow access to all rows, columns and diagonals, to enable fast rendering and parallel manipulations of volumetric structures [COB93]. The solution for the diagonal Latin cube that we give here is presented in [COK95]. The idea is to have a mapping from each element (x, y, z) in a cube to one of the n memory modules. Such mappings are also called *skewing schemes* because they skew the trivial mapping from element to memory.

Diagonal Latin cube

Let us first consider the simpler problem where we ignore conflicts along diagonals and consider only a Latin cube. For that, the following simple skewing function maps (x, y, z) onto the mth

0	1	2	3	4
2	3	4	0	1
4	0	1	2	3
1	2	3	4	0
3	4	0	1	2

0	1	2	3	4	5	6	7
3	4	5	6	7	0	1	2
6	7	0	1	2	3	4	5
1	2	3	4	5	6	7	0
4	5	6	7	0	1	2	3
7	0	1	2	3	4	5	6
2	3	4	5	6	7	0	1
5	6	7	0	1	2	3	4

(a)   (b)

Figure 13.5: Two examples of diagonal Latin squares.

memory module:

$$m = x + y + z \pmod{n}, \quad 0 <= x, y, z, \ m < n.$$

Since in each row or column along any dimension, two coordinates remain constant; this skewing function guarantees that, for any value of the free variable, a different memory module number is generated. Note that this solution works for any n, and that this skewing function is not unique, and there are many other functions (as we shall see later) that can generate such Latin cubes.

Now, let us also consider the diagonals, that is, along all rows, columns and diagonals, each module index appears only once. Two skewing functions that generate such diagonal Latin squares are displayed in Figure 13.5: (a) $m = x + 2y \pmod{5}$ and (b) $m = 2x + 6y \pmod{7}$. These are linear skewing schemes of the form,

$$m = Ax + By + Cz \pmod{N} \quad 0 <= x, y, z < n,$$

where the coefficients A, B and C define the properties of the mapping. In the following we will see which conditions these coefficients should meet to guarantee conflict-free access to rows, column or diagonals. Note that we immediately discuss the 3D case; in 2D, simply $C \equiv 0$.

Rows along the X-axis are conflict-free if A and n are *coprime* or *relatively prime*, which means that their greatest common divisor (gcd) is 1. Similarly, rows along the Y-axis or the Z-axis, B or

C, respectively, must be relatively prime to n to be conflict-free. Let's denote this as

$$\gcd(A, n) = 1 \,, \quad \gcd(B, n) = 1 \,, \quad \gcd(C, n) = 1 \,.$$

Clearly, when n is prime, no matter which coefficients we use, we get no conflict along the axis. For a standard chessboard (where $n = 8$), however, there must be an odd number of coefficients so no two entries will be mapped to the same module.

Now, to guarantee conflict-free diagonals requires us to consider more than one axis or all the possible combinations of two or three coefficients. First, let us consider diagonals within an orthogonal slice of a cube, where only two axes change and one remains constant. We consider only the directions of these two-axis diagonals (ignoring their signs). The sum or difference of the two acting coefficients should be relatively prime to n to be conflict-free:

$$\gcd(A \pm B, n) = 1 \,, \quad \gcd(B \pm C, n) = 1 \,, \quad \gcd(A \pm C, n) = 1 \,.$$

Note that even if n is prime, the sum of two coefficients can still be equal to n and conflicts will occur. Note also that for $n = 8$, since the coefficients must be odd, their difference is not, and conflict along the diagonals cannot be avoided (see Figure 13.3).

Similarly, diagonals that traverse along the three axes (also known as the major diagonals of a cube), require the combination of the three coefficients:

$$\gcd(A \pm B \pm C, n) = 1 \,.$$

Note that all the above 13 conditions on A, B and C are unoriented, since their signs do not change the gcd.

The reader can easily verify that the skewing functions that generated the diagonal Latin squares in Figure 13.5 satisfy the necessary conditions presented above. More details and a 3D example can be found in [COK95].

More conflict-free access

Skewing schemes have great importance in parallel-computer architecture, since they allow simultaneous access to several structures in vector multiprocessors. Also, when these accesses are

performed in an asynchronous manner, these schemes can lower
the number of collisions in the network and the conflicts in memory modules, avoiding high latencies. There is a very rich research
literature on this topic. Here we will just show that the above
linear skewing schemes can also be effective in allowing conflict-
free access to blocks, for example, a block of 4×4 pixels in an
image. Such simple schemes can be useful in the implementation
of various image space algorithms on the graphics processing unit
(GPU) or any multiprocessor or interleaved memories. For example, see the implementation of a voxel-based fly-through on a
multiprocessor [CORLS96].

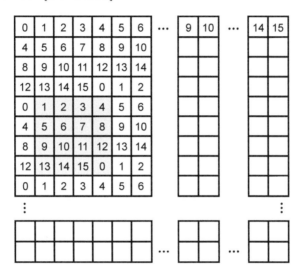

Figure 13.6: A skewing scheme that allows conflict-free access to 4×4 blocks.

Figure 13.6 illustrates the linear skewed scheme generated by

$$m = x + 4y \;(\text{mod}\,16)\,.$$

The access to any 4×4 block is conflict-free. Note that the illustration highlights one such block. Such conflict-free access to
a 2D structure can prove to be effective for applying a filter or a
convolution with another 4×4 block, for example. Note, however,
that this skewing scheme is not conflict-free along row or columns,
since 4 is not relatively prime to 16.

Figure 13.7 illustrates a similar linear skewed scheme generated
by

$$m = x + 4y \;(\text{mod}\,17)\,.$$

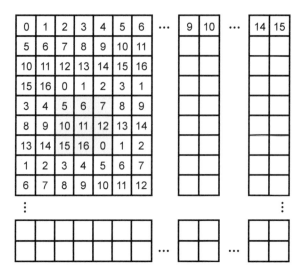

Figure 13.7: A diagonal Latin square that allows conflict-free access to 3×3 blocks.

Here, it uses a prime number, which results in a diagonal Latin square. However, there are no free lunches, and this scheme enables conflict-free access to a 3×3 block only. Note that in a 4×4 block there are only two conflicts though.

It should be noted that the cost of modulo 16 (or any 2^k number) operation is much cheaper than that of a prime number, since it can be implemented by a simple bitwise shift operation. Sometimes, such considerations are important enough for accelerating the calculations of computationally intensive tasks.

Bibliography

[ACCO05]   Jackie Assa, Yaron Caspi, and Daniel Cohen-Or, *Action synopsis: Pose selection and illustration*, ACM Trans. Graph. **24** (2005), no. 3, 667–676.

[ACOT⁺10]   O. K.-C. Au, D. Cohen-Or, C.-L. Tai, H. Fu, and Y. Zheng, *Electors voting for fast automatic shape correspondence*, Computer Graphics Forum **29** (2010), no. 2, 645–654.

[ACSD⁺03]   Pierre Alliez, David Cohen-Steiner, Olivier Devillers, Bruno Lévy, and Mathieu Desbrun, *Anisotropic polygonal remeshing*, ACM Trans. Graph. **22** (2003), no. 3, 485–493.

[Aga07]   Aseem Agarwala, *Efficient gradient-domain compositing using quadtrees*, ACM Trans. Graph. **26** (2007), no. 3, 94.

[AMD02]   Pierre Alliez, Mark Meyer, and Mathieu Desbrun, *Interactive Geometry Remeshing*, ACM Trans. Graph. **21(3)** (2002), 347–354.

[Ami02]   I. Amidror, *Scattered data interpolation methods for electronic imaging systems: A survey*, Journal of Electronic Imaging **11** (2002), no. 2, 157–76.

[AS07]   Shai Avidan and Ariel Shamir, *Seam carving for content-aware image resizing*, ACM Trans. Graph. **26** (2007), no. 3, 10.

[BCM99]   Richard K. Beatson, Jon B. Cherrie, and C. T. Mouat, *Fast fitting of radial basis functions: Methods based on preconditioned gmres iteration*, Advances in Computational Mathematics **11** (1999), 253–270.

[BDR+06]   Yoshua Bengio, Olivier Delalleau, Nicolas Le Roux, Jean-Francois Paiement, Pascal Vincent, and Marie Ouimet, *Spectral dimensionality reduction*, Feature Extraction, Foundations and Applications, Springer, 2006.

[BGSF08]   S. Biasotti, D. Giorgi, M. Spagnuolo, and B. Falcidieno, *Reeb graphs for shape analysis and applications*, Theor. Comput. Sci. **392** (2008), no. 1–3, 5–22.

[BK04]   Yuri Boykov and Vladimir Kolmogorov, *An experimental comparison of min-cut/max-flow algorithms for energy minimization in vision*, IEEE Trans. Pat. Ana. & Mach. Int. (PAMI) **26** (2004), no. 9, 1124–1137.

[BL97]   R. K. Beatson and W. A. Light, *Fast evaluation of radial basis functions: Methods for two-dimensional polyharmonic splines*, IMA Journal of Numerical Analysis **17** (1997), 343–372.

[BM93]   Serge Beucher and Fernand Meyer, *The morphological approach to segmentation: The watershed transformation*, Mathematical Morphology in Image Processing (E.R. Dougherty, ed.), Marcel Dekker Inc., 1993, pp. 433–481.

[Bra99]   Ronald N. Bracewell, *The Fourier transform and its applications*, McGraw-Hill, 1999.

[BS08]   Mario Botsch and Olga Sorkine, *On linear variational surface deformation methods*, IEEE Trans. Vis. & Comp. Graphics **14** (2008), no. 1, 213–230.

[BV99]   Volker Blanz and Thomas Vetter, *A morphable model for the synthesis of 3D faces*, Proc. ACM SIGGRAPH, 1999, pp. 187–194.

[BV06]   Yuri Boykov and Olga Veksler, *Handbook of mathematical models in computer vision*, ch. Graph Cuts in Vision and Graphics: Theories and Applications, Springer, 2006.

[CBC+01]   J. C. Carr, R. K. Beatson, J. B. Cherrie, T. J.
           Mitchell, W. R. Fright, B. C. McCallum, and T. R.
           Evans, *Reconstruction and representation of 3D ob-
           jects with radial basis functions*, Proc. ACM SIG-
           GRAPH, 2001, pp. 67–76.

[CC94]   Trevor F. Cox and Michael A. A. Cox, *Multidimen-
           sional scaling*, Chapman & Hall, 1994.

[CGNT09]   C. Couprie, L. Grady, L. Najman, and H. Talbot,
           *Power watersheds: A new image segmentation frame-
           work extending graph cuts, random walker and op-
           timal spanning forest*, Proc. Int. Conf. Comp. Vis.
           (ICCV), 2009, pp. 731–738.

[COB93]   Daniel Cohen-Or and Reuven Bakalash, *The con-
           veyor: An interconnection device for parallel vol-
           umetric transformations*, Rendering, Visualization
           and Rasterization Hardware (Eurographics'91 Work-
           shop), 1993, pp. 77–85.

[COK95]   Daniel Cohen-Or and Arie Kaufman, *A 3D skewing
           and de-skewing scheme for conflict-free access to rays
           in volume rendering*, IEEE Transactions on Comput-
           ers **44** (1995), no. 5, 707–710.

[CORLS96] D. Cohen-Or, E. Rich, U. Lerner, and V. Shenkar,
           A real-time photo-realistic visual flythrough, IEEE
           Trans. Vis. & Comp. Graphics **2** (1996), no. 3, 255–
           265.

[CSRL01]   Thomas H. Cormen, Clifford Stein, Ronald L. Rivest,
           and Charles E. Leiserson, *Introduction to algorithms*,
           2nd ed., McGraw-Hill Higher Education, 2001.

[Dav06]   T. A. Davis, *Direct methods for sparse linear systems*,
           SIAM, 2006.

[dC76]   Manfredo P. do Carmo, *Differential geometry of
           curves and surfaces*, Prentice-Hall, 1976.

[EK03]   A. Elad and R. Kimmel, *On bending invariant signa-
           tures for surfaces*, IEEE Trans. Pat. Ana. & Mach.
           Int. (PAMI) **25** (2003), no. 10, 1285–1295.

[EY36]   C. Eckart and G. Young, *The approximation of one matrix by another of lower rank*, Psychometrika **1** (1936), 211–218.

[FF62]   L. R. Ford and D. R. Fulkerson, *Flows in networks*, Princeton University Press, Princeton, NJ, 1962.

[FHL⁺09]   Zeev Farbman, Gil Hoffer, Yaron Lipman, Daniel Cohen-Or, and Dani Lischinski, *Coordinates for instant image cloning*, ACM Trans. Graph. **28** (2009), no. 3, 67:1–67:9.

[FN80]   R. Franke and G. M. Nielson, *Scattered data interpolation of large sets of scattered data*, International Journal of Numerical Methods in Engineering **15** (1980), 1691–1704.

[GCO06]   Ran Gal and Daniel Cohen-Or, *Salient geometric features for partial shape matching and similarity*, ACM Trans. Graph. **25** (2006), no. 1, 130–150.

[GPS89]   D. Greig, B. Porteous, and A. Seheult, *Exact maximum a posteriori estimation for binary images*, Journal of the Royal Statistical Society Series B **51** (1989), 271–279.

[GS00]   I. M. Gelfand and R. A. Silverman, *Calculus of variations*, Dover Publications, 2000.

[HLMS04]   Jihun Ham, Daniel D. Lee, Sebastian Mika, and Bernhard Schölkopf, *A kernel view of the dimensionality reduction of manifolds*, Proc. Int. Conf. Machine learning (ICML), 2004, pp. 47–54.

[Jai89]   A. K. Jain, *Fundamentals of digital image processing*, Prentice Hall, 1989.

[JLCW06]   Zhongping Ji, Ligang Liu, Zhonggui Chen, and Guojin Wang, *Easy mesh cutting*, Computer Graphics Forum **25** (2006), no. 3, 283–291.

[Joh97]   Andrew Johnson, *Spin-images: A representation for 3-D surface matching*, Ph.D. thesis, Robotics Institute, Carnegie Mellon University, Pittsburgh, PA, August 1997.

[JSTS06]   Jiaya Jia, Jian Sun, Chi-Keung Tang, and Heung-Yeung Shum, *Drag-and-drop pasting*, ACM Trans. Graph. **25** (2006), no. 3, 631–637.

[JZH07]   Tao Ju, Qian-Yi Zhou, and Shi-Min Hu, *Editing the topology of 3D models by sketching*, ACM Trans. Graph. **26** (2007), no. 3, 42:1–42:9.

[JZvK07]   Varun Jain, Hao Zhang, and Oliver van Kaick, *Non-rigid spectral correspondence of triangle meshes*, Int. J. Shape Modeling **13** (2007), no. 1, 101–124.

[KG00]   Z. Karni and C. Gotsman, *Spectral compression of mesh geometry*, Proc. ACM SIGGRAPH, 2000, pp. 279–286.

[KR05]   Byung Moon Kim and Jarek Rossignac, *Geofilter: Geometric selection of mesh filter parameters*, Computer Graphics Forum **24** (2005), no. 3, 295–302.

[KSE+03]   Vivek Kwatra, Arno Schödl, Irfan Essa, Greg Turk, and Aaron Bobick, *Graphcut textures: Image and video synthesis using graph cuts*, ACM Trans. Graph. **22** (2003), no. 3, 277–286.

[KT03]   S. Katz and A. Tal, *Hierarchical mesh decomposition using fuzzy clustering and cuts*, ACM Trans. Graph. **22** (2003), no. 3, 954–961.

[KvD92]   Jan J. Koenderink and Andrea J. van Doorn, *Surface shape and curvature scales*, Image Vision Comput. **10** (1992), no. 8, 557–565.

[LDB05]   Guillaume Lavoué, Florent Dupont, and Atilla Baskurt, *A new cad mesh segmentation method, based on curvature tensor analysis*, Computer Aided Design **37** (2005), no. 10, 975–987.

[Low99]   David Lowe, *Object recognition from local scale-invariant features*, Proceedings of the International Conference on Computer Vision, Volume 2, 1999, pp. 1150–1157.

[LSTS04]   Yin Li, Jian Sun, Chi-Keung Tang, and Heung-Yeung Shum, *Lazy snapping*, ACM Trans. Graph. **23** (2004), 303–308.

[MC08]   F. C. Monteiro and A. Campilho, *Watershed framework to region-based image segmentation*, Proc. Int. Conf. Pat. Rec. (ICPR), 2008, pp. 1–4.

[MGP06]   N. J. Mitra, L. Guibas, and M. Pauly, *Partial and approximate symmetry detection for 3D geometry*, ACM Trans. Graph. **25** (2006), no. 3, 560–568.

[MP00]   Geoffrey McLachlan and David Peel, *Finite mixture models*, Wiley Series in Probability and Statistics, Wiley, 2000.

[MPWC12]   Niloy J. Mitra, Mark Pauly, Michael Wand, and Duygu Ceylan, *Symmetry in 3D geometry: Extraction and applications*, EUROGRAPHICS State-of-the-art Report, 2012.

[OBS03]   Yutaka Ohtake, Alexander Belyaev, and Hans-Peter Seidel, *A multi-scale approach to 3D scattered data interpolation with compactly supported basis functions*, Proc. Int. Conf. Shape Modeling and Applications (SMI), 2003, pp. 153–161.

[PGB03]   Patrick Pérez, Michel Gangnet, and Andrew Blake, *Poisson image editing*, ACM Trans. Graph. **22** (2003), no. 3, 313–318.

[PKA03]   D. L. Page, A. F. Koschan, and M. A. Abidi, *Perception-based 3D triangle mesh segmentation using fast marching watersheds*, Proc. IEEE Conf. Comp. Vis. Pat. Rec. (CVPR), 2003, pp. 27–32.

[PP93]   U. Pinkall and K. Polthier, *Computing discrete minimal surfaces and their conjugates*, Experimental Mathematics **2** (1993), no. 1, 15–36.

[PSBM07]   Valerio Pascucci, Giorgio Scorzelli, Peer-Timo Bremer, and Ajith Mascarenhas, *Robust on-line computation of Reeb graphs: Simplicity and speed*, ACM Trans. Graph. **26** (2007), no. 3, 58.

[RKB04]   Carsten Rother, Vladimir Kolmogorov, and Andrew Blake, *"GrabCut": Interactive foreground extraction using iterated graph cuts*, ACM Trans. Graph. **23** (2004), 309–314.

[RSA08]   Michael Rubinstein, Ariel Shamir, and Shai Avidan, *Improved seam carving for video retargeting*, ACM Trans. Graph. **27** (2008), no. 3, 16:1–16:9.

[RZ04]   L. Roditty and U. Zwick, *On dynamic shortest paths problems*, In Proceedings of 12th Annual European Symposium on Algorithms (ESA), 2004, pp. 580–591.

[SA09]   Ariel Shamir and Shai Avidan, *Seam carving for media retargeting*, Communications of the ACM **52** (2009), no. 1, 77–85.

[Saa03]   Y. Saad, *Iterative methods for sparse linear systems*, Second Edition, SIAM, 2003.

[SCO04]   Olga Sorkine and Daniel Cohen-Or, *Least-squares meshes*, Proc. Int. Conf. Shape Modeling and Applications (SMI), 2004, pp. 191–199.

[She68]   Donald Shepard, *A two-dimensional interpolation function for irregularly-spaced data*, Proc. ACM National Conference, 1968, pp. 517–524.

[SLS+07]   Andrei Sharf, Thomas Lewiner, Gil Shklarski, Sivan Toledo, and Daniel Cohen-Or, *Interactive topology-aware surface reconstruction*, ACM Trans. Graph. **26** (2007), no. 3, 43.

[SS97]   Richard Szeliski and Heung-Yeung Shum, *Creating full view panoramic image mosaics and environment maps*, Proc. ACM SIGGRAPH, 1997, pp. 251–258.

[SS09]   Ariel Shamir and Olga Sorkine, *Visual media retargeting*, ACM SIGGRAPH ASIA 2009 Courses, SIGGRAPH ASIA '09, 2009, pp. 11:1–11:13.

[SSCO08]   Lior Shapira, Ariel Shamir, and Daniel Cohen-Or, *Consistent mesh partitioning and skeletonization using the shape diameter function*, The Visual Computer **24** (2008), no. 4, 249–259.

[SSGH01]   P. Sander, J. Snyder, S. Gortler, and H. Hoppe, *Texture mapping progressive meshes*, Proc. ACM SIGGRAPH, 2001, pp. 409–416.

[SSM97]   Bernhard Schölkopf, Alex J. Smola, and Klaus-Robert Müller, *Kernel principal component analysis*, Proc. Int. Conf. Artificial Neural Networks, 1997, pp. 583–588.

[SSSE00]   Arno Schödl, Richard Szeliski, David H. Salesin, and Irfan Essa, *Video textures*, Proc. ACM SIGGRAPH, 2000, pp. 489–498.

[STC04]   John Shawe-Taylor and Nello Cristianini, *Kernel methods for pattern analysis*, Cambridge University Press, 2004.

[Tau95]   G. Taubin, *A signal processing approach to fair surface design*, Proc. of ACM SIGGRAPH, 1995, pp. 351–358.

[TB97]   Lloyd N. Trefethen and David Bau, *Numerical linear algebra*, SIAM, 1997.

[TP91]   M. Turk and A. Pentland, *Eigenfaces for recognition*, Journal of Cognitive Neuroscience **3** (1991), no. 1, 71–86.

[TV04]   Johan Tangelder and Remco Veltkamp, *A survey of content based 3D shape retrieval methods*, Proc. Int. Conf. Shape Modeling and Applications (SMI), 2004, pp. 145–156.

[VH99]   R. Veltkamp and M. Hagedoorn, *State-of-the-art in shape matching*, Tech. Report UU-CS-1999-27, Utrecht University, the Netherlands, 1999.

[VL08]   B. Vallet and B. Lévy, *Spectral geometry processing with manifold harmonics*, Computer Graphics Forum **27** (2008), no. 2, 251–260.

[VS91]   Luc Vincent and Pierre Soille, *Watersheds in digital spaces: An efficient algorithm based on immersion simulations*, IEEE Trans. Pat. Ana. & Mach. Int. (PAMI) **13** (1991), no. 6, 583–598.

[WHDS04]   Zoë Wood, Hugues Hoppe, Mathieu Desbrun, and
            Peter Schröder, *Removing excess topology from iso-
            surfaces*, ACM Trans. Graph. **23** (2004), no. 2, 190–
            208.

[WL97]   K. Wu and M. D. Levine, *3D part segmentation using
            simulated electrical charge distributions*, IEEE Trans.
            Pat. Ana. & Mach. Int. (PAMI) **19** (1997), no. 11,
            1223–1235.

[ZH04]   Yinan Zhou and Zhiyong Huang, *Decomposing poly-
            gon meshes by means of critical points*, Proc. In-
            ternational Multimedia Modelling Conference, 2004,
            pp. 187–195.

[ZKK02]   G. Zigelman, R. Kimmel, and N. Kiryati, *Texture
            mapping using surface flattening via multidimen-
            sional scaling*, IEEE Trans. Vis. & Comp. Graphics
            8 (2002), no. 2, 198–207.

[ZvKD10]   Hao Zhang, Oliver van Kaick, and Ramsay Dyer,
            Spectral mesh processing, Computer Graphics Forum
            29 (2010), no. 6, 1865–1894.

Index

T - #0518 - 071024 - C244 - 254/178/11 - PB - 9780367658786 - Gloss Lamination